Lecture Notes in Mathematics 2030

For further volumes:
http://www.springer.com/series/304

Yukio Matsumoto
José María Montesinos-Amilibia

Pseudo-periodic Maps and Degeneration of Riemann Surfaces

Springer

Yukio Matsumoto
Gakushuin University
Department of Mathematics
Mejiro 1-5-1
171-8588 Toshima-ku Tokyo
Japan
yukiomat@math.gakushuin.ac.jp

José María Montesinos-Amilibia
Universidad Complutense
Facultad de Matemáticas
Departamento de Geometría
y Topología
Plaza de las Ciencias 3
28040 Madrid
Spain
montesin@mat.ucm.es

ISBN 978-3-642-22533-8 e-ISBN 978-3-642-22534-5
DOI 10.1007/978-3-642-22534-5
Springer Heidelberg Dordrecht London New York

Lecture Notes in Mathematics ISSN print edition: 0075-8434
ISSN electronic edition: 1617-9692

Library of Congress Control Number: 2011934808

Mathematics Subject Classification (2010): 14-XX, 57-XX

Cover design: deblik, Berlin

Printed on acid-free paper

Springer is part of Springer Science+Business Media (www.springer.com)

Dedicated with respect and affection to the memory of Professor Itiro Tamura (1926–1991)

Preface

In 1944, Nielsen introduced a certain type of mapping classes of a surface which were called by him *surface transformation classes of algebraically finite type*, [53]. He introduced this type of mapping classes as a generalization of the mapping classes of finite order. By the celebrated Nielsen Theorem [52], the latter classes contain surface homeomorphisms of finite order (For a generalization, see Kerckhoff [30]). A mapping class of algebraically finite type does not necessarily contain a homeomorphism of finite order, but using Nielsen's theorem [52], one can show that it contains a homeomorphism f satisfying the following conditions (in what follows f will be an orientation-preserving homeomorphism of a closed, connected, oriented surface of genus g, Σ_g):

1. There exists a disjoint union of simple closed curves (which will be called *cut curves*)
$$\mathscr{C} = C_1 \cup C_2 \cup \cdots \cup C_r$$
on Σ_g such that $f(\mathscr{C}) = \mathscr{C}$, and
2. the restriction of f to the complement of $\mathscr{C}$,
$$f|(\Sigma_g - \mathscr{C}) : \Sigma_g - \mathscr{C} \to \Sigma_g - \mathscr{C}$$

is isotopic to a periodic map, namely a homeomorphism of finite order. (Cf. [53, Sect. 14], [22]).

In the present memoir, such a homeomorphism (and also a homeomorphism which is isotopic to such a homeomorphism) will be called a *pseudo-periodic map*. A periodic map is a special case of a pseudo-periodic map. In recent terminology, a homeomorphism f is pseudo-periodic if and only if either it is of finite order or its mapping class $[f]$ is reducible and all the component mapping classes are of finite order. (See [12,16,22,24,63]). A surface transformation class of algebraically finite type is nothing but a mapping class of a pseudo-periodic map.

Nielsen [53] studied these classes extensively and defined several important invariants, for instance, the *screw number* of f about a cut curve C_i which measures the amount of the (fractional) Dehn-twist performed by a certain power f^α of f sending C_i to itself; or the character of C_i: whether it is "amphidrome" or not. Here C_i is *amphidrome* if there is an integer γ such that

$$f^\gamma(\vec{C_i}) = -\vec{C_i}.$$

He asserted in [53] that his invariants were a complete set of conjugacy invariants, meaning that if two pseudo-periodic maps

$$f_1 : \Sigma_g^{(1)} \to \Sigma_g^{(1)}$$

and

$$f_2 : \Sigma_g^{(2)} \to \Sigma_g^{(2)}$$

have these same invariants, then their mapping classes $[f_1]$ and $[f_2]$ are equivalent under a certain homeomorphism

$$h : \Sigma_g^{(1)} \to \Sigma_g^{(2)},$$

i.e. $[f_1] = [h^{-1} f_2 h]$. (For an exact formulation, see [22, Theorem 13.4]). However, his proof of this assertion was rather vague, and we need an invariant (the action of monodromy on the partition graph) which he did not mention explicitly. See Examples 6.3 and 6.4 in Chap. 6.

A pseudo-periodic map f is said to be *of negative twist* if the screw numbers about a certain system of cut curves are all negative (Chap. 3). The purpose of Part I of the present memoir is to construct a complete set of conjugacy invariants for a pseudo-periodic map f of negative twist.

We have added to Nielsen's invariants one more: *the action of f on the "partition graph"*, which is the action, induced by f, on the configuration of the partition of Σ_g obtained by cutting Σ_g along a certain system of cut curves $\{C_i\}_{i=1}^r$. The main result of Part I is roughly stated as follows (see Theorem 6.1 and 6.3 for precise statements):

Theorem 0.1. *Let $f_1 : \Sigma_g^{(1)} \to \Sigma_g^{(1)}$ and $f_2 : \Sigma_g^{(2)} \to \Sigma_g^{(2)}$ be pseudo-periodic maps of negative twist. Suppose that they have the same values in Nielsen's invariants and that their actions on the respective partition graphs are equivariantly isomorphic. Then there exists an orientation preserving homeomorphism $h : \Sigma_g^{(1)} \to \Sigma_g^{(2)}$ such that $[f_1] = [h^{-1} f_2 h]$.*

In the course of the proof, we develop (in Chaps. 3–5) the theory of *generalized quotients*, which are naturally associated with pseudo-periodic maps, just as ordinary quotient spaces are associated with periodic maps. This makes our proof of Theorem 0.1 unexpectedly long, but the generalized quotients will play an essential role also in the study of the degeneration of Riemann surfaces (in Part II). This was the main reason of our investigation, which therefore concentrated in the study of generalized quotients. As a matter of fact, Theorem 0.1 above is just a (non immediate) corollary of our research.

The organization of Part I is as follows:

In Chap. 1, we review some basic results of Nielsen from [51, 53].

In Chap. 2, we define the "standard form" of a pseudo-periodic map f. Nielsen [53, Sect. 14] constructed a special homeomorphism which served as a standard form, but our standard form is slightly different from his. We prove the existence and the essential uniqueness of the homeomorphism in standard form which is isotopic to a given pseudo-periodic map (Theorem 2.1).

In Chap. 3, we introduce the notion of generalized quotients, and in particular, of *minimal quotients* which are the special case of generalized quotients that satisfy a certain "minimality condition". According to the definition given in Chap. 3, in order to have a generalized quotient, a pseudo-periodic map f must be in a very special form which we would like to call "superstandard form". It will be proved that any pseudo-periodic map f of negative twist is isotopic to a pseudo-periodic map in superstandard form having a minimal quotient (Theorem 3.1).

In Chap. 4, the following essential uniqueness will be proved (Theorem 4.1): suppose f_1 and $f_2 : \Sigma_g \to \Sigma_g$ are pseudo-periodic maps of negative twist, both in superstandard form. If they are homotopic, then their respective minimal quotients are isomorphic.

By the above existence and uniqueness theorems, we can generalize the definition of minimal quotients to cover any pseudo-periodic map of negative twist not necessarily in superstandard form, i.e., the minimal quotient of a pseudo-periodic map f of negative twist is constructed by first isotoping f to the superstandard form f' and then taking the minimal quotient of f', which is declared to be the minimal quotient of f.

The minimal quotient captures all of the Nielsen invariants constructed in [53]. Moreover, it will be proved in Part II that the "base space" of the minimal quotient of a pseudo-periodic map f of negative twist is homeomorphic to a (normally minimal) singular fiber of a one-parameter family of Riemann surfaces of genus g around which the topological monodromy is equivalent to $[f]$.

In Chap. 5, we prove a theorem in elementary number theory, which is basic to the arguments in Chaps. 3 and 4.

In Chap. 6, we consider the partition graph and the action of f on it. This action, together with the minimal quotient, determines the conjugacy class of $[f]$ in $\mathscr{M}_g$(Theorem 6.1). This result is further reformulated in terms of certain cohomology of "weighted graphs" (Theorem 6.3).

In Appendix A, we will give a proof of the following theorem: *let f and f' be (orientation-preserving) periodic maps of a compact surface Σ each component of which has negative Euler characteristic. Suppose f and $f' : (\Sigma, \partial\Sigma) \to (\Sigma, \partial\Sigma)$ are homotopic as maps of pairs. Then there exists a homeomorphism $h : \Sigma \to \Sigma$ isotopic to the identity, such that $f = h^{-1} f' h$.*

This theorem is used in the proof of Theorem 2.1. Among specialists, this theorem seems folklore. A. Edmonds informed, in a letter to the second named author, that C. Frohman had proved a stronger result which implied the above theorem. Unfortunately, the authors could not find any reference giving an explicit proof, so we decided to write this appendix.

Pseudo-periodic maps of negative twist are closely related to the degeneration of Riemann surfaces. In fact, the topological monodromy around a singular fiber in a one-parameter family of Riemann surfaces is a pseudo-periodic map of negative twist (see [19], also [26, 58]).

In Part II of this memoir, we will apply the results in Part I to the topology of degeneration of Riemann surfaces. The main result of Part II is roughly summarized as follows:

Theorem 0.2. *The topological types of minimal degenerating families of Riemann surfaces of genus $g \geq 2$, over a disk, which are nonsingular outside the origin, are in a bijective correspondence with the conjugacy classes in the mapping class group $\mathcal{M}_g$ represented by pseudo-periodic maps of negative twist. The correspondence is given by the topological monodromy.*

In the case of $g = 1$, the validity of Theorem 0.2 is reduced by half: By Kodaira's classification [32] of singular fibers for genus 1, we see that every pseudo-periodic mapping class (of negative twist) of a torus can be realized as the topological monodromy of a singular fiber. Thus the correspondence is "surjective", but it is not "injective". For example, all the multiple fibers of type ${}_m\mathrm{I}_0$ (in Kodaira's notation) have the identity mapping class as their topological monodromy.

The assumption $g \geq 2$ is used almost everywhere in our proof: The existence of an admissible system of cut curves subordinate to a pseudo-periodic map (Lemma 1.1) is essential to the definition of various invariants of the map, and the proof of the existence requires $g \geq 2$. Also "homotopy implies conjugacy" theorem for periodic maps assumes $g \geq 2$, because in the proof we apply the hyperbolic geometry (see Appendix A). This theorem is indispensable in the proof of the uniqueness of the standard form (see Theorem 2.1 (ii)). Our arguments in later chapters depend on this uniqueness theorem.

We have tried to make the memoir as self-contained as possible, except for the two quotations from [51, 53]. (Theorems 1.1 and 1.2 of the present memoir). All the other arguments are elementary.

The authors are grateful to Allan Edmonds, Takayuki Oda and Hiroshige Shiga for their useful information and comments.

This work started during the first named author's first visit to Spain (1988) and was completed during his second visit (1991). The first named author would like to express his warmest thanks to the members and staffs of Facultad de Ciencias Matemáticas, Universidad Complutense de Madrid, for their kind hospitality.

Finally but not least at all, the authors deeply thank Srta. María Angeles Bringas for her benevolence, patience, and excellent skill shown in typing this memoir, without which it could have never been published.

ADDED:SEPTEMBER,2009[1]

The main body of the manuscript of the present memoir was completed in December 1991, and some remaining additional parts in January, 1992. Since then we have not found any occasion to publish this work, for several reasons; the length certainly was one. Another reason, but probably the main one, was the authors' inability to use Tex.

After a long delay of almost two decades, the authors find some unsatisfactory points in the manuscript, for example, it contains too many details, which might be a hindrance for readers who want to get a quick view, but on the other hand, it might help them to understand the details of the argument. Anyway the authors needed to compile these long (elementary and sometimes seemingly trivial) arguments to complete the proof of our theorems. Therefore, we have decided to keep the manuscript in its original form, except for the numbering of the chapters, theorems, propositions, figures, etc. A change we have made is the unification of the two different bibliographies, which were separately attached to each part, into one bibliography at the end. Also we added some references that were published after the completion of our manuscript and some more that we had missed involuntarily or were unknown to us due to our limitations. Unfortunately, the authors cannot be sure even now of the completeness of the augmented bibliography.

A pseudo-periodic map would well be called *chiral* if either it is periodic or all of its screw numbers are of the same sign. A chiral pseudo-periodic map is a pseudo-periodic map of negative twist or of positive twist. If a pseudo-periodic map has both positive and negative screw numbers, it will be called *achiral*. In Part I of this memoir, we confined ourselves to chiral pseudo-periodic maps (of negative twist). From the viewpoint of surface topology, it would be more natural to treat not only chiral pseudo-periodic maps but also achiral ones, of course. We tried such a general treatment for some time. However, to construct a generalized quotient for an achiral pseudo-periodic map, we are forced to adopt an artificial convention on signs of intersections between the components consisting of a *tail*-part of the generalized quotient, and we lose the *natural* uniqueness of the generalized quotient of a pseudo-periodic map. Moreover, our construction of generalized quotients is intended to be applied to topology of degeneration of Riemann surfaces in Part II, and for that purpose, we only need chiral pseudo-periodic maps. For these reasons, we gave up our trial to generalize the construction of generalized quotients to achiral maps.

As is immediately seen from the title, the main objects of this memoir are (chiral) pseudo-periodic homeomorphisms and degeneration of Riemann surfaces. Our main point is that these two objects are topologically classified by the same objects, i.e. certain types of "numerical chorizo spaces" together with a cohomology class in the weighted cohomology of their decomposition graphs. This type of chorizo spaces appear as "minimal quotients"of pseudo-periodic homeomorphisms of negative twist, and exactly the same type of chorizo spaces appear also as "normaly minimal

[1]Revised in February 2011.

singular fibers" in one-parameter families of Riemann surfaces. The former objects come from surface topology, while the latter objects come from complex analysis. The authors think interesting that numerical chorizo spaces lie at the common basis of the objects from two different disciplines.

The appearance of pseudo-periodic homeomorphisms of negative twist in degenerating families of Riemann surfaces was clarified through the work of Imayoshi [26], Shiga–H. Tanigawa [58], and finally by Earle and Sipe [19]. We should mention here, however, that the pseudo-periodic nature of the monodromy had been observed for the Milnor fibering [44] at an isolated singular point of a complex hypersurface.

Brieskorn [14] showed, in general dimensions, that the eigenvalues of the (co)homological monodromy are roots of unity. Lê [34] showed, in the case of curves, that the homological monodromy is periodic if the curve is irreducible at the singular point. A'Campo [1] proved that it is not the case if the curve is not irreducible. Also he showed that the geometric (i.e. topological) monodromy is not necessarily periodic, even if the curve is irreducible. A'Campo [2] and Eisenbud – Neumann [20] gave a description of geometric pseudo-periodic monodromies. Finally Lê – Michel – Weber [35] proved that the geometric monodromy is pseudo-periodic ("quasi-finie" in their terminology). Michel – Weber [43] gave a detailed description of the negative twist and showed that the geometric monodromy associated to a complex polynomial map from $\mathbb{C}^2$ to $\mathbb{C}$ (affine case) is also pseudo-periodic of negative twist.

During the two decades, after the completion of our manuscript, several related papers have appeared.

The most related one is, of course, the anouncement of this memoir, which was published in Bull. A.M.S. in 1994 [42]. This might serve as an introduction to this memoir (see also [40]). Pichon [55] used the pseudo-periodicity of the geometric monodromy to characterize the 3-manifolds that appear as the boundary manifolds of degenerating families of Riemann surfaces over a disk. In both of the papers of Pichon [55] and Lê – Michel – Weber [35], Waldhausen's graph manifolds [66, 67] play an important role.

The first authors that put the present memoir to good use were Ashikaga and Ishizaka [7] who gave a complete list of singular fibers in degenerating families of genus 3 (they were more than sixteen hundred!). They very explicitely exploited the algorithm, implicitly contained in the present memoir. It should be noted that the numerical classification of genus 3 singular fibers had been accomplished by Uematsu [64] in 1993 independently of our work.

Xiao and Reid [56] proposed the problem of determining all the "atomic" singular fibers, which are defined as such singular fibers that cannot be "split" by any perturbation of the degenerating families. This problem is very interesting from the viewpoint of the present memoir. By our main result, the topological types of singular fibers are classified by the corresponding topological monodromies around them. Then a natural question to be settled would be if all atomic singular fibers (except for "multiple fibers") correspond to the full (-1)-Dehn twist about a certain simple closed curve. Examples of this geometrical situation are contained in [3, 29, 38, 39]. (For recent related results, see [5, 6, 8].) Following these trend of ideas, S.Takamura [59, 60] is undertaking a project in solving this problem.

The authors would like to thank Professors D. T. Lê, J.-P. Brasselet, and M. Oka who showed interest in our work, and especially Professor Lê for his explanation on the related results of his own and others. We are also grateful to Professor T. Ashikaga for taking our work seriously and for actively developing related subjects in algebraic geometry and topology, which encouraged us very much. Thanks are also due to Professor Y. Imayoshi for his interest in our results, and for his very benevolent review of our work [28].

In November of 2000 we met in Oberwolfach Professors A'Campo, Weber and Pichon, who encouraged us very strongly to publish our results. We owe to them the final impulse that we needed to conclude the typing of this memoir that has eventually lead to its publication.

Tokyo and Madrid,
September 2009

Yukio Matsumoto
José María Montesinos-Amilibia

Acknowledgement

The authors are very grateful to the referees for their careful reading, valuable comments and suggestions. The first named author has been supported by Grant-in-Aid for Scientific Research (No. 20340014), J.S.P.S. The second named author has been supported by MEC, MTM2009-07030.

Contents

Part I
Conjugacy Classification of Pseudo-periodic Mapping Classes

In Part I, we will study pseudo-periodic maps and will give a complete set of conjugacy invariants of *chiral* pseudo-periodic mapping classes. The main theorems of Part I are Theorems 6.1 and 6.3.

Chapter 1
Pseudo-periodic Maps

1.1 Basic Definitions

In this Chapter, we will review some basic results from Nielsen [51, 53].

We will begin with the definition of a pseudo-periodic map. Hereafter, all surfaces will be *oriented*, and all homeomorphisms between them will be *orientation-preserving*.

Let Σ_g be a closed connected surface of genus $g \geq 1$.

Definition 1.1 (Compare [53, Sect. 8]). A homeomorphism

$$f : \Sigma_g \to \Sigma_g$$

is a *pseudo-periodic map* if f is isotopic to a homeomorphism

$$f' : \Sigma_g \to \Sigma_g$$

which satisfies the following conditions:

(i) there exists a disjoint union of simple closed curves

$$\mathscr{C} = C_1 \cup C_2 \cup \cdots \cup C_r$$

on Σ_g such that $f'(\mathscr{C}) = \mathscr{C}$, and

(ii) the restriction

$$f'| \left(\Sigma_g - \mathscr{C}\right) : \Sigma_g - \mathscr{C} \to \Sigma_g - \mathscr{C}$$

of f' to the complement of $\mathscr{C}$, is isotopic to a periodic map, i.e. a mapping of finite order.

We call $\{C_i\}_{i=1}^r$ a *system of cut curves* subordinate to f. It may be empty if f is isotopic to a periodic map.

It is easy to see that a homeomorphism of a torus

$$f : T^2 \to T^2$$

Y. Matsumoto and J.M. Montesinos-Amilibia, *Pseudo-periodic Maps and Degeneration of Riemann Surfaces*, Lecture Notes in Mathematics 2030,
DOI 10.1007/978-3-642-22534-5_1, © Springer-Verlag Berlin Heidelberg 2011

is pseudo-periodic if and only if $|Trace\ (f_*)| \leq 2$, where

$$f_* : H_1(T^2; \mathbf{Z}) \to H_1(T^2; \mathbf{Z})$$

is the induced homomorphism.

In what follows, we will assume $g \geq 2$ unless otherwise stated.

Lemma 1.1. *Let*

$$f : \Sigma_g \to \Sigma_g$$

be a pseudo-periodic map. Then there exists a system of cut curves $\{C_i\}_{i=1}^r$ *subordinate to* f *such that*

(i) C_i *does not bound a disk on* Σ_g, $i = 1, 2, \ldots, r$, *and*
(ii) C_i *is not parallel to* C_j *if* $i \neq j$.

Proof. Let $\{C_j{}'\}_{j=1}^s$ be any system of cut curves subordinate to f. Isotoping f, we may assume

$$f\left(\bigcup_j C_j{}'\right) = \bigcup_j C_j{}'.$$

If a cut curve $C_j{}'$ bounds a disk, we simply remove $C_j{}'$. Suppose that $C_i{}'$ and $C_j{}'$ bounds an annulus A. Since $g \geq 2$, A is uniquely determined by $C_i{}'$ and $C_j{}'$. Let m be the smallest positive integer such that $f^m(A) = A$. We equivariantly isotop f on

$$A \cup f(A) \cup \cdots \cup f^{m-1}(A)$$

so that f^m maps the "center line" C_A of A onto itself. Then we omit $C_i{}'$ and $C_j{}'$ and also their images under the iteration of f, and add instead, C_A together with its images. Two curves $C_i{}'$ and $C_j{}'$ are replaced by a curve C_A.

Proceeding in this way, we arrive at a system of cut curves $\{C_i\}_{i=1}^r$ satisfying (i) and (ii) of the lemma. □

We call a system of cut curves which satisfies conditions (i) and (ii) of Lemma 1.1 an *admissible system* ([12, 22]).

Let $\{C_i\}_{i=1}^r$ be an admissible system of cut curves subordinate to a pseudo-periodic map

$$f : \Sigma_g \to \Sigma_g.$$

Then Nielsen [53, Sect. 12] introduced the "screw number" $s(C_i)$ for each curve C_i in the system. The definition is as follows:

We may assume

$$f\left(\bigcup_i C_i\right) = \bigcup_i C_i.$$

Let C_j be a fixed cut curve in the the system.

Let α be the smallest positive integer α such that

$$f^{\alpha}(\vec{C}_j) = \vec{C}_j$$

where $\vec{C}_j$ denotes C_j with an orientation assigned, so f^{α} preserves Let b and b' be the connected components of $\Sigma_g - \mathscr{C}$ such that C_j belongs to their adherence in Σ_g, where

$$\mathscr{C} = \bigcup_{i=1}^{r} C_i.$$

It is possible that $b = b'$. Let β (resp. β') be the smallest positive integer such that $f^{\beta}(b) = b$ (resp. $f^{\beta'}(b') = b'$). Clearly α is a common multiple of β and β'.

Since $f|\left(\Sigma_g - \mathscr{C}\right)$ is isotopic to a periodic map, there exists a positive integer n_b such that $(f^{\beta}|b)^{n_b} \cong \mathrm{id}_b$. We choose the smallest number among such integers and denote it by n_b again. Likewise we choose $n_{b'}(> 0)$ for b'.

Let L be the least common multiple of $n_b\beta$ and $n_{b'}\beta'$.

Then f^L is isotopic to the identity on b and b'. Thus, on the union

$$b \cup C_j{}' \cup b',$$

f^L is isotopic to the result of a number of full Dehn twists performed about C_j. Let $e(\in \mathbf{Z})$ be this number of full twists.

We adopt the following convention for the sign of e:

Convention (†). The sign convention of e is depicted in Fig. 1.1. (Compare [13, p. 166], where the orientation of the surface is opposite to ours; and compare [46, Fig. 3 and p. 158]).

Since the system $\{C_i\}_{i=1}^r$ is admissible, neither b nor b' is an annulus or a disk. Nielsen [53] proved that the number e is well-defined.

Definition 1.2 ([53, Sect. 12]). The rational number $e\alpha/L$ is called the *screw number* of f about C_j and is denoted by $s(C_j)$. It measures the "amount of Dehn-twist" performed by f^{α} about C_j.

If $s(C_j) = 0$, we can equivariantly isotop f on

$$b \cup C_j \cup b'$$

and on its images under the iteration of f so that the result of the isotopy is a periodic map of

$$b \cup C_j \cup b'.$$

Thus we can omit C_j, together with its images under the iteration of f, keeping the remaining system admissible.

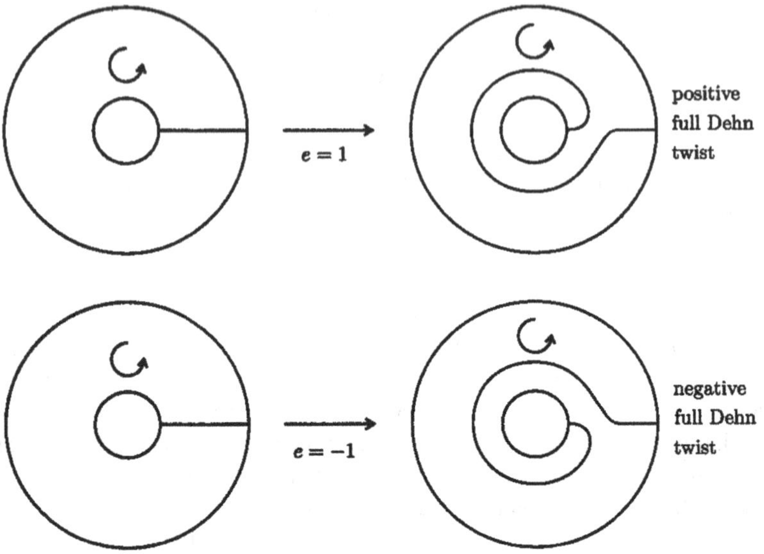

Fig. 1.1 Positive and negative Dehn twists

Iterating this omission, we arrive at a cut system in which all cut curves have non-zero screw number.

Definition 1.3 (Cf. [12,22]). An admissible system of cut curves $\{C_i\}_{i=1}^r$ is *precise* if $s(C_i) \neq 0$ for each C_i.

Nielsen did not use the terminology "precise system"; he talked, instead, about "division of the surface into complete kernels" ([53, Sect. 13]) meaning the decomposition of the surface by a precise system.

Definition 1.4 ([53, Sect. 10]). An essential simple closed curve C on Σ_g is said to be *amphidrome* with respect to a homeomorphism

$$f : \Sigma_g \to \Sigma_g$$

if there is an integer γ such that $f^\gamma(\overrightarrow{C})$ is freely homotopic to $-\overrightarrow{C}$, where $\overrightarrow{C}$ and $-\overrightarrow{C}$ denote the same C with the opposite directions assigned.

1.2 Nielsen's Results on Pseudo-periodic Maps

The main result of Nielsen's paper [53] is the following:

Theorem 1.1 ([53, Sect. 15], cf. [22, p. 90]). *Let*

$$f : \Sigma_g \to \Sigma_g$$

and

$$f' : \Sigma_g \to \Sigma_g$$

be pseudo-periodic maps, and let $\{C_i\}_{i=1}^r$ *and* $\{C_j{}'\}_{j=1}^s$ *be precise systems of cut curves subordinate to* f *and* f' *respectively. Suppose* f *is homotopic to* f'. *Then*

(i) *there is a homeomorphism*

$$h : \Sigma_g \to \Sigma_g$$

which is isotopic to the identity and such that $h(\mathscr{C}) = \mathscr{C}'$, *where*

$$\mathscr{C} = \bigcup_{i=1}^{r} C_i$$

and

$$\mathscr{C}' = \bigcup_{i=1}^{s} C_j{}'.$$

In particular, $r = s$.

(ii) *For each cut curve* C_i *in* $\{C_i\}_{i=1}^r$, *the number* α *and the screw number* $s(C_i)$ *are equal to the corresponding numbers for* $h(C_i)$.

(iii) *For each cut curve* C_i *in* $\{C_i\}_{i=1}^r$, *its character of being amphidrome or not (with respect to* f*) coincides with the same character of* $h(C_i)$ *(with respect to* f'*).*

(iv) *For each connected component* b *of* $\Sigma_g - \mathscr{C}$, *the integers* β *and* n_b *are equal to the corresponding integers for* $h(b)$.

(v) *For each connected component* b *of* $\Sigma_g - \mathscr{C}$, *the periodic map of* b *which is isotopic to* $f^\beta|b$ *is conjugate to the periodic map of* $h(b)$ *which is isotopic to*

$$(f')^\beta|h(b).$$

(Here we are assuming $f(\mathscr{C}) = \mathscr{C}$ and $f'(\mathscr{C}') = \mathscr{C}$, which can be achieved by isotoping f and f' if necessary).

Nielsen [53, Sect. 15] also asserted that conversely these invariants determine the conjugacy class of the mapping class to which f belongs. This, however, is not the case. In fact, the converse as formulated as Theorem 13.4 of [22] has counter examples. See Sect. 7, Examples, 3 and 4, of the present paper.

If in the statement of Theorem 1.1 we add the following condition:

(vi) the actions of f and f' on the respective partition graphs are conjugate (Here the *partition graph* of (Σ_g, f) is a graph whose vertices and edges correspond to connected components of $\Sigma_g - \mathscr{C}$ and cut curves $\{C_i\}_{i=1}^r$, respectively. Similarly for (Σ_g, f')), then Nielsen's assertion is correct at least for pseudo-periodic maps of negative twists (see Sect. 4 for the definition). The conditions (i)–(vi) become sufficient. See Sect. 7 for details.

The part of Nielsen's result stated above is certainly correct, because he developed his theory so that it depends only on the homotopy class of f. For the construction of h in assertion (i), consult, for example, Casson's lecture notes ([16, Lemma 2.4]). Note that by the above theorem we can speak of the precise system of cut curves subordinate to f, provided that we identify two isotopic systems up to indexing of the curves.

As is obvious from assertion (v), we must study periodic maps as a prerequisite to Theorem 1.1. Nielsen [51] studied periodic maps of surfaces and established a complete set of conjugacy invariants, which we will now describe.

Let Σ be a compact surface with or without boundary,

$$f : \Sigma \to \Sigma$$

a periodic map of order $n > 1$.

Let p be a point of Σ. Then there is a positive integer $m = m(p)$ such that the points

$$p,\ f(p),\ \dots f^{m-1}(p)$$

are distinct and $f^m(p) = p$. If $m = n$, the point p is called a *simple point*, while if $0 < m < n$, p is called a *multiple points*. In particular, if $m = 1$, p is a *fixed point*. Since we are assuming that f is orientation-preserving, a multiple point is an isolated, interior point of Σ.

For later use, we consider a more general situation than is needed here. Let

$$\vec{\mathscr{C}} = \vec{C}_1 \cup \vec{C}_2 \cup \cdots \cup \vec{C}_s$$

be a set of oriented and disjoint simple closed curves in a surface Σ, and let g be a map $g : \Sigma \to \Sigma$ such that $g(\vec{\mathscr{C}}) = \vec{\mathscr{C}}$ and $g|\vec{\mathscr{C}}$ is periodic. We will define the notion of the *valency* of a curve $\vec{C} \in \vec{\mathscr{C}}$ *with respect to* g.

Let m be the smallest positive integer such that $g^m(\vec{C}) = \vec{C}$. The restriction $g^m|\vec{C}$ is a periodic map of $\vec{C}$ of order, say, $\lambda > 0$. Then $m\lambda = n$ (= the order of $g|\vec{C}$). Let q be any point on $\vec{C}$, and suppose that the images of q under the iteration of g^m are situated on $\vec{C}$ in this order.

$$\{q,\ g^{\sigma m}(q),\ g^{2\sigma m}(q),\ \dots,\ g^{(\lambda-1)\sigma m}(q)\}$$

when viewed in the direction of $\vec{C}$. Here, by Convention (†), the integer σ satisfies $0 \le \sigma < \lambda$ and $gcd(\sigma, \lambda) = 1$, so $\sigma = 0$ iff $\lambda = 1$. Define an integer δ as follows:

$$\sigma\delta \equiv 1 \pmod{\lambda}, \quad 0 \le \delta < \lambda$$

(NB. $\delta = 0$ iff $\lambda = 1$). Then the action of g^m on $\vec{C}$ is topologically equivalent to the rotation of angle $2\pi\delta/\lambda$ in a circle. If λ is known, σ and δ give the same information.

Definition 1.5 ([51, Sect. 2]). The triple (m, λ, σ) is called the *valency* of $\vec{C}$ *with respect to* g.

If the orientation of C is reversed, then m and λ do not change, while $\sigma \pmod{\lambda}$ changes sign.

Returning to our original periodic map f of order n acting on a surface Σ, we define the *valency of a boundary curve* as its valency (with respect to f) assuming it has the orientation induced by the surface Σ. The *valency of a multiple point* p is defined to be the valency of the boundary curve $\partial D_{p'}$ orientated from the *inside* of an invariant disk neighborhood D_p.

The valency (m, λ, σ) of a curve C (with respect to f) is common to its images

$$f(C),\ f^2(C),\ \dots,\ f^{m-1}(C),$$

so this notion descends to the quotient space Σ/f.

The quotient space Σ/f is an *orbifold* ([62], cf. [57]). Its underlying space is a compact surface. A multiple point $p \in \Sigma$ corresponds to a *cone point* $\in \Sigma/f$. Thus we can speak of the *valency of a cone point* of Σ/f. Also we can speak of the *valency of a boundary curve* of Σ/f.

Theorem 1.2 ([51, Sect. 11]). *Let*

$$f : \Sigma \to \Sigma$$

and

$$f' : \Sigma' \to \Sigma'$$

be periodic maps of the same order of compact, connected, mutually homeomorphic surfaces Σ *and* Σ'. *Then* f *and* f' *are conjugate if and only if one can establish a bijective correspondence between the set of the boundary curves and the cone points of* Σ/f *and the set of the boundary curves and the cone points of* Σ'/f', *respectively, so that corresponding boundary curves (and cone points) have the same valency.*

We remark that $f : \Sigma \to \Sigma$ and $f' : \Sigma' \to \Sigma'$ are *conjugate* if, by definition, there is an *orientation-preserving* homeomorphism $h : \Sigma \to \Sigma'$ such that $f = h^{-1} f' h$. (Beware that Nielsen's original statement [51, Sect. 11] is a little confusing with regard to orientation, but correct).

Corollary 1.1 ([51, Sect. 11]). *Let*

$$f : \Sigma \to \Sigma$$

and

$$f' : \Sigma' \to \Sigma'$$

be periodic maps of the same order, of compact connected mutually homeomorphic surfaces Σ *and* Σ'. *If* Σ *and* Σ' *are closed surfaces, and* f *and* f' *have no multiple points, then* f *and* f' *are conjugate.*

Theorem 1.2, Corollary 1.1, and Theorem 1.3 below hold regardless of the genus of the surfaces.

Example 1.1 ([51, Sect. 6]). Parametrize a torus T^2 by $\mathbf{R}^2/\mathbf{Z}^2$. Define periodic maps of order 5

$$f_1, f_2 : T^2 \to T^2$$

as follows:

$$f_1(x, y) = (x, y + 1/5),$$
$$f_2(x, y) = (x, y + 2/5).$$

Then f_1 and f_2 are conjugate. In fact, a homeomorphism $h : T^2 \to T^2$ defined by

$$h(x, y) = (3x + 5y, 4x + 7y)$$

gives the conjugation:

$$f_1 = h^{-1} f_2 h.$$

For later use (especially in Sect. 7), we need a somewhat more detailed version of Theorem 1.2.

Theorem 1.3. *Let*

$$f : \Sigma \to \Sigma$$

and

$$f' : \Sigma' \to \Sigma'$$

be periodic maps of the same order, of compact connected, mutually homeomorphic surfaces Σ *and* Σ'. *Suppose they have no multiple points. Let* $M = \Sigma/f$ *and* $M' = \Sigma'/f'$ *be the respective quotient spaces, and suppose there is a homeomorphism*

$$H : M \to M'$$

which preserves the valencies of the boundary curves. Then there exists another homeomorphism

$$H' : M \to M'$$

and a homeomorphism

$$h : \Sigma \to \Sigma'$$

such that

(i) the diagram

$$\begin{array}{ccc} \Sigma & \xrightarrow{h} & \Sigma' \\ {\scriptstyle f}\downarrow & & \downarrow{\scriptstyle f'} \\ \Sigma & \xrightarrow{h} & \Sigma' \\ {\scriptstyle \pi}\downarrow & & \downarrow{\scriptstyle \pi'} \\ M & \xrightarrow{H'} & M' \end{array}$$

commutes: (π, π' *being the natural projections*)
(ii) $H'|\partial M = H|\partial M$.

This theorem is essentially proved by Nielsen [51], but our version is slightly stronger. We now give an outline of the proof. Let n be the order of the periodic maps f, f'.

First we recall the definition of the *monodromy exponent* ω from [51] (which was denoted by μ in [51, Sect. 2]. Let

$$l : [0, 1] \to M$$

be a loop, $l(0) = l(1)$. Let

$$\tilde{l} : [0, 1] \to \Sigma$$

be a lift. Then there is an integer $k \pmod n$ such that

$$f^k(\tilde{l}(0)) = \tilde{l}(1),$$

and $\omega(l)$ is defined to be $k \pmod n$. Nielsen calls the mod n integer $\omega(l)$ the *monodromy exponent* of the (directed) loop l, which depends only on the homology class of l. We will call the homomorphism

$$\omega : H_1(M) \to \mathbf{Z}/n$$

the *monodromy exponent*, too.

Suppose M has genus q and s boundary components. Choosing a base point O on M, Nielsen draws a canonical system of $2q + s$ curves

$$\alpha_1, \beta_1, \ldots, \alpha_q, \beta_q, \gamma_1, \ldots, \gamma_s$$

on M. (Fig. 1.2).

Then he defines eight special homeomorphism (called by him "generating mappings")

Fig. 1.2 Figure 1 of [51]

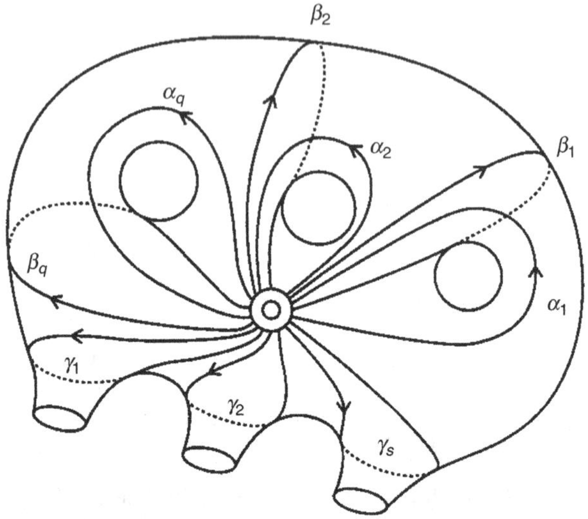

$$S_1,\ \ldots,\ S_8 : M \to M.$$

He properly chooses a combination of generating mappings, so that when it is applied to the original system of curves, a new canonical system (again denoted by $\alpha_1, \beta_1, \ldots, \alpha_q, \beta_q, \gamma_1, \ldots, \gamma_s$) satisfying the following conditions

$$\begin{aligned} \omega(\alpha_1) &= 1 \\ \omega(\alpha_i) &= 0,\ 2 \le i \le q, \\ \omega(\beta_i) &= 0,\ 1 \le i \le q, \end{aligned}$$

is obtained. (See [51, Sect. 5]).

Suppose we have drawn on M' a canonical system of $2q + s$ curves

$$\alpha_1{}',\ \beta_1{}',\ \ldots,\ \alpha_q{}',\ \beta_q{}',\ \gamma_1{}',\ \ldots,\ \gamma_s{}'$$

with the same properties as above.

Recall (in our version) that we are given a homeomorphism

$$H : M \to M'$$

which preserves the valencies of the boundary curves. This H does not in general preserve the above canonical systems, of course.

We will construct a new homeomorphism

$$H' : M \to M'$$

satisfying

(A) H' maps the canonical system on M to the one on M', and
(B) $H'|\partial M = H|\partial M$.

We must be careful here because it is impossible in general to construct such a homeomorphism *unless* we change properly the choice of the loops $\gamma_1', \dots, \gamma_s'$.

Let

$$\Delta_s(\subset M)$$

be a disk with s holes which contains $\gamma_1, \gamma_2, \dots, \gamma_s$, and such that $M - \Delta_s$ is a connected surface (with one boundary component) containing

$$\alpha_1,\ \beta_1,\ \dots,\ \alpha_q,\ \beta_q.$$

Since

$$(\alpha_1,\ \beta_1,\ \dots,\ \alpha_q,\ \beta_q,\ \gamma_1,\ \dots,\ \gamma_s)$$

is a cannonical system, such a Δ_s exists. (See Fig. 1.3). We take a similar disk with s holes Δ_s' in M', which contains

$$\gamma_1',\ \gamma_2',\ \dots,\ \gamma_s'$$

and does not contain

$$\alpha_1',\ \beta_1',\ \dots,\ \alpha_q',\ \beta_q'.$$

Suppose the loops

$$\gamma_1,\ \gamma_2,\ \dots,\ \gamma_s$$

are attached to the base point O on M in this order (as in Fig. 1.3). Let Γ_i be the boundary curve of M around which the loop γ_i goes, $i = 1, 2, \dots, s$. We denote the image $H(\Gamma_i)$ by Γ_i'.

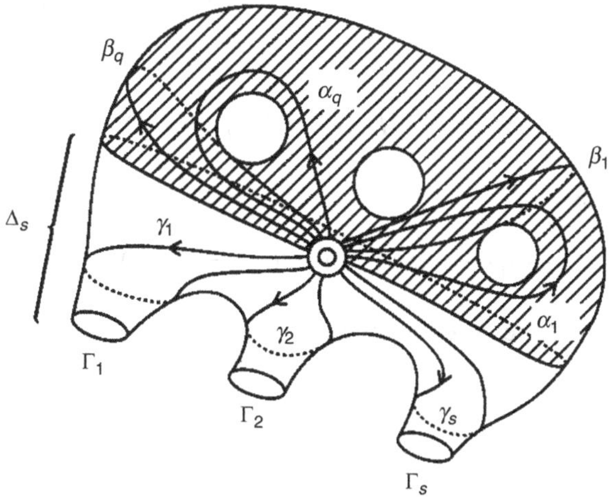

Fig. 1.3 A disk with s holes Δ_s

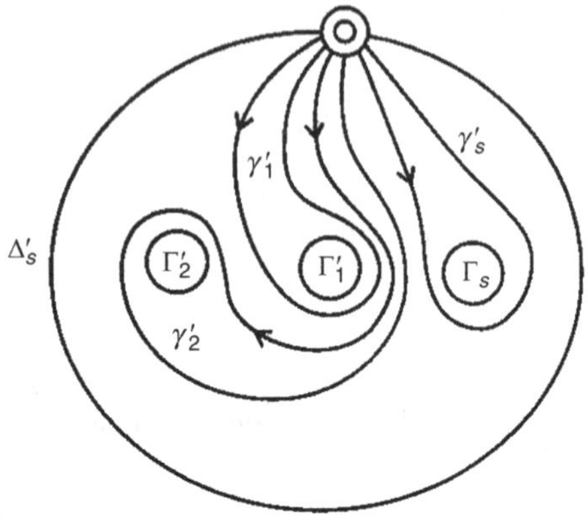

Fig. 1.4 Rechosen loops $\gamma'_1, \gamma'_2, \ldots, \gamma'_s$

Beware that $\gamma_i{}'$ does not necessarily go around the boundary curve $\Gamma_i{}' = (H\Gamma_i)$. We must re-choose $\gamma_i{}'$ as follows: look at the disk with s holes $\Delta_s{}'$. We re-choose

$$\gamma_1{}',\ \gamma_2{}',\ \ldots,\ \gamma_s{}'$$

inside $\Delta_s{}'$ so that

(1) $\gamma_i{}'$ goes around $\Gamma_i{}'$, and
(2)

$$\gamma_1{}',\ \gamma_2{}',\ \ldots,\ \gamma_s{}'$$

are attached to O' in this order. (See Fig. 1.4)

Now we can define a homeomorphism

$$H' : M \to M'$$

such that

$$H'(\alpha_i) = \alpha_i{}',\ H'(\beta_i) = \beta_i{}',\ H'(\gamma_j) = \gamma_j{}'$$

and

$$H'|\partial M = H|\partial M.\ (i.e.\ H'(\Gamma_j) = \Gamma_j{}'),\quad i = 1,\ \ldots,\ q;\ j = 1,\ \ldots,\ s.$$

Since

$$\omega(\alpha_1) = \omega(\alpha_1{}') = 1,\ \omega(\alpha_2) = \omega(\alpha_2{}') = 0, \ldots,\ \omega(\beta_q) = \omega(\beta_q{}') = 0$$

and H preserves the valencies of the boundary curves, we infer that

$$H' : M \to M'$$

preserves the monodromy exponents of curves. Now follow Nielsen's argument in [51, Sect. 11] and construct a homeomorphism

$$h : \Sigma \to \Sigma'$$

such that $f = h^{-1} f' h$ and $\pi' h = H' \pi$. This completes the outline of the proof of Theorem 1.3. □

Chapter 2
Standard Form

Given a pseudo-periodic map

$$f : \Sigma_g \to \Sigma_g$$

Nielsen constructed a special homeomorphism which is homotopic to f and plays the role of a "standard form" in the mapping class of f, [53, Sect. 14]. In this Chapter, we will construct a similar standard form, slightly different from Nielsen's, and will show its essential uniqueness. He wanted to avoid fixed points which might appear in annular neighborhoods of cut curves, while we do not care about such fixed points (Compare [22, Theorem 13.3]).

2.1 Definitions and Main Theorem of Chap. 2

Definition 2.1. Let A be an annulus and

$$\phi : [0, 1] \times S^1 \to A$$

a parametrization (i.e. homeomorphism), where $S^1 = \mathbf{R}/\mathbf{Z}$. A homeomorphism

$$f : A \to A$$

which does not interchange the boundary components of A is called a *linear twist with respect to* ϕ, if

$$f\phi(t, x) = \phi(t, x + at + b), (t, x) \in [0, 1] \times S^1,$$

for some $a, b \in \mathbf{Q}$. We say simply that

$$f : A \to A$$

Y. Matsumoto and J.M. Montesinos-Amilibia, *Pseudo-periodic Maps and Degeneration of Riemann Surfaces*, Lecture Notes in Mathematics 2030,
DOI 10.1007/978-3-642-22534-5_2,

is a *linear twist* if f is a linear twist with respect to a certain parametrization

$$\phi : [0,1] \times S^1 \to A.$$

Definition 2.2. Let A and

$$\phi : [0,1] \times S^1 \to A$$

be as above. A homeomorphism

$$f : A \to A$$

which interchanges the boundary components of A is called a *special (piecewise-linear) twist with respect to* ϕ, if

$$f\phi(t,x) = \begin{cases} \phi\left(1-t, -x-3a\left(t-\frac{1}{3}\right)\right), & 0 \le t \le \frac{1}{3} \\ \phi(1-t, -x), & \frac{1}{3} \le t \le \frac{2}{3} \\ \phi\left(1-t, -x-3a\left(t-\frac{2}{3}\right)\right), & \frac{2}{3} \le t \le 1 \end{cases}$$

for some $a \in \mathbf{Q}$. If

$$f : A \to A$$

is a special twist with respect to a certain parametrization

$$\phi : [0,1] \times S^1 \to A,$$

we simply say that f is a *special twist*.

Remark 2.1. Let

$$\rho : [0,1] \times S^1 \to [0,1] \times S^1$$

be defined by

$$\rho(t,x) = \rho(1-t, -x).$$

If

$$f : A \to A$$

is a special twist with respect to

$$\phi : [0,1] \times S^1 \to A,$$

then this same f is also a special twist with respect to

$$\phi\rho : [0,1] \times S^1 \to A.$$

Now we define our standard form.

Definition 2.3. *A pseudo-periodic map*

$$f : \Sigma_g \to \Sigma_g$$

is said to be in *standard form* if the following conditions are satisfied:

(i) There exists a system of disjoint annular neighborhoods $\{A_i\}_{i=1}^r$ of the precise system of cut curves subordinate to f, such that

$$f(\mathscr{A}) = \mathscr{A},$$

where

$$\mathscr{A} = \bigcup_{i=1}^{r} A_i.$$

(ii) The map

$$f \mid \Sigma_g - \mathscr{A} : \Sigma_g - \mathscr{A} \to \Sigma_g - \mathscr{A}$$

is periodic.

(iii) Let k_i be the smallest positive integer such that

$$f^{k_i}(A_i) = A_i, i = 1, 2, \ldots, r.$$

(iii)-a If

$$f^{k_i} \mid A_i : A_i \to A_i$$

does not interchange the boundary components of A_i, then $f^{k_i} \mid A_i$ is a linear twist.

(iii)-b If

$$f^{k_i} \mid A_i : A_i \to A_i$$

interchanges the boundary components of A_i, then $f^{k_i} \mid A_i$ is a special twist.

This whole chapter is devoted to the proof of the following theorem.

Theorem 2.1 (cf. [53, Sect. 15] [22, Theorem 13.3]). *(i) Any pseudo-periodic map*

$$f : \Sigma_g \to \Sigma_g$$

is isotopic to a pseudo-periodic map in standard form.

(ii) Suppose two pseudo-periodic maps in standard form

$$f, f' : \Sigma_g \to \Sigma_g$$

are homotopic, then there is a homeomorphism

$$h : \Sigma_g \to \Sigma_g$$

isotopic to the identity, such that $f = h^{-1} f' h$.

The uniqueness statement (ii) above is stronger than the original one ([53, Sect. 15]).

The proof requires considerations on periodic parts, non-amphidrome annuli, and amphidrome annuli. These preliminary considerations occupy the main body of this chapter.

2.2 Periodic Part

We need the following theorem.

Theorem 2.2. *Let f and f' be periodic maps of a compact surface Σ each component of which has negative Euler characteristic. Suppose*

$$f, f' : (\Sigma, \partial\Sigma) \to (\Sigma, \partial\Sigma)$$

are homotopic as maps of pairs. Then there exists a homeomorphism $h : \Sigma \to \Sigma$ isotopic to the identity, such that $f = h^{-1} f' h$.

This theorem seems folklore among specialists. A. Edmonds informed, in a letter to the second named author, that C. Frohman had proved a stronger result in his thesis which implied Theorem 2.2. Unfortunately the authors could not find any reference giving an explicit proof. A proof will be given in Appendix A.

2.3 Non-amphidrome Annuli

Lemma 2.1 (Linearization). *Let A be an annulus,*

$$f : A \to A$$

a homeomorphism which does not interchange the boundary components. Suppose

$$f \mid \partial A : \partial A \to \partial A$$

is periodic. Then there exists an isotopy

$$f_\tau : A \to A, \quad 0 \leq \tau \leq 1,$$

such that

$$f_0 = f, f_\tau \mid \partial A = f \mid \partial A, and$$

$f_1 : A \to A$ is a linear twist. (An isotopy as this will be referred to hereafter as *"rel.∂").*

Proof. Let $\partial_0 A$ and $\partial_1 A$ denote the two boundary components. Since

$$f \mid \partial A : \partial A \to \partial A$$

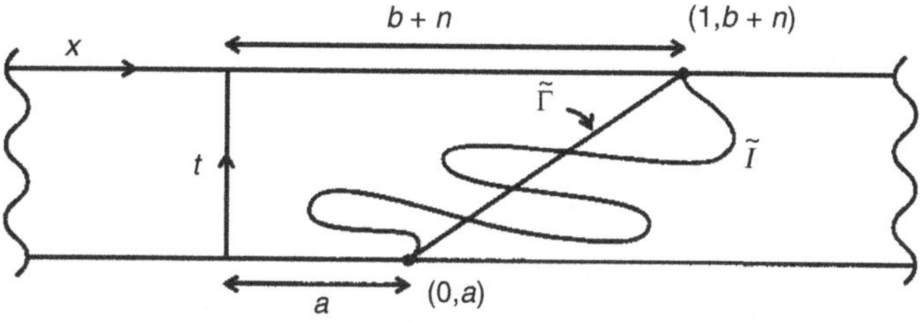

Fig. 2.1 Linearization

is periodic, there are homeomorphisms

$$\varphi : S^1 \to \partial_0 A$$

and

$$\psi : S^1 \to \partial_1 A$$

such that

$$f\varphi(x) = \varphi(x + a), f\psi(x) = \psi(x + b),$$

for $a, b \in \mathbf{Q}$.

Note that φ and ψ are homotopic as maps from S^1 into A. Thus there is a homeomorphism

$$\phi : [0, 1] \times S^1 \to A$$

such that

$$\phi(0, x) = \varphi(x), \phi(1, x) = \psi(x)$$

([21, Lemma 2.4]). Now we identify A with $[0, 1] \times S^1$ through ϕ. We lift the curve $f([0, 1] \times \{0\})$ to the universal covering $[0, 1] \times \mathbf{R}$. Let $\tilde{I}$ be the lift starting at $(0, a)$. Then it ends at $(1, b + n)$, for some $n \in \mathbf{Z}$. (See Fig. 2.1)

Let $\tilde{\Gamma}$ be a line segment in $[0, 1] \times \mathbf{R}$ joining $(0, a)$ and $(1, b + n)$. Then $\tilde{\Gamma}$ projects to a "linear" arc Γ in $[0, 1] \times S^1$ connecting $(0, a)$ to $(1, b)$. By an innermost arc argument, f is rel.∂ isotopic to a homeomorphism

$$f' : [0, 1] \times S^1 \to [0, 1] \times S^1$$

satisfying

$$f'([0, 1] \times \{0\}) = \Gamma.$$

Also by composing f' with an isotopy which keeps ∂A pointwise fixed and moves along Γ, f' becomes isotopic to a homeomorphism $f^{(2)}$ such that

$$f^{(2)}(t, 0) = (t, *) \in \Gamma,$$

i.e., $f^{(2)}$ preserves the t-level when restricted to a map from $[0,1]\times\{0\}$ onto Γ.

Define

$$f^{(3)}:[0,1]\times S^1\to[0,1]\times S^1$$

by

$$f^{(3)}(t,x)=(t,x+a(1-t)+(b+n)t).$$

$f^{(3)}$ is a linear twist which coincides with $f^{(2)}$ on $\partial A\cup\Gamma$. Thus by the Alexander trick, $f^{(2)}$ is isotopic to $f^{(3)}$ keeping the points of $\partial A\cup\Gamma$ fixed. □

It will be convenient to extend the notion of screw number of a curve to an annulus. Let

$$\mathscr{A}=\bigcup_{i=1}^{r}A_i$$

be a disjoint union of annuli $f:\mathscr{A}\to\mathscr{A}$ a homeomorphism. Suppose the restriction of f to the boundary

$$\partial\mathscr{A}=\bigcup_{i=1}^{r}\partial A_i$$

is periodic. Let A_i be an annulus in $\mathscr{A}$. Let α be the smallest positive integer such that (i) $f^{\alpha}(A_i)=A_i$; and (ii) f^{α} does not interchange the boundary components. Let l be a non-zero integer such that $f^{l}\mid\partial A_i=$ the identity. Then l is a multiple of α, and $f^{l}:A_i\to A_i$ is the result of e full Dehn-twists, e being an integer.

Definition 2.4. The rational number $e\alpha/l$ is called the *screw number* of f in A_i and is denoted by $s(A_i)$. It measures the amount of Dehn twist performed by f^{α} in A_i.

The number $s(A_i)$ is independent of the choice of l. Of course if $\{A_i\}_{i=1}^{r}$ is an invariant system of annular neighborhoods of a precise cut system $\{C_i\}_{i=1}^{r}$ subordinate to a pseudo-periodic map

$$f:\Sigma_g\to\Sigma_{g'}$$

then

$$s(A_i)=s(C_i),\quad i=1,2,\dots,r.$$

An annulus A_i is said to be *amphidrome* if there is an integer γ such that $f^{\gamma}(A_i)=A_i$ and f^{γ} interchanges the boundary components.

The *valency* (m,λ,σ) of a boundary curve $\vec{C}$ of $\mathscr{A}$ oriented by the orientation induced from $\mathscr{A}$ is defined as the valency of $-\vec{C}$ with respect to the periodic map

$$\partial\mathscr{A}\to\partial\mathscr{A}.$$

As a corollary to (the proof of) Lemma 2.1, we have the following:

Corollary 2.1. *Let* $\mathscr{A} = \bigcup_{i=1}^{r} A_i$ *be a disjoint union of annuli,*

$$f : \mathscr{A} \to \mathscr{A}$$

a homeomorphism such

$$f \mid \partial\mathscr{A} \to \partial\mathscr{A}$$

is periodic. Let A_i *be a non-amphidrome annulus of* $\mathscr{A}$ *with*

$$\partial A_i = \partial_0 A_i \cup \partial_1 A_1.$$

Let

$$(m_i^0, \lambda_i^0, \sigma_i^0)$$

and

$$(m_i^1, \lambda_i^1, \sigma_i^1)$$

be the valencies of $\partial_0 A_i$*, and* $\partial_1 A_i$*, respectively,* $s(A_i)$ *the screw number. Then*

(i) *the equality*

$$m_i^0 = m_i^1$$

holds; and

(ii) *the number*

$$s(A_i) + \delta_i^0/\lambda_i^0 + \delta_i^1/\lambda_i^1$$

is an integer, where the integer δ_i^ν *is determined by*

$$\sigma_i^\nu \delta_i^\nu \equiv 1 \pmod{\lambda_i^\nu}$$

and

$$0 \leq \delta_i^\nu < \lambda_i^\nu, \quad \nu = 0, 1.$$

Proof. The oriented A_i will be denoted by $\overrightarrow{A_i}$; the induced orientation in ∂A_i by $\partial \overrightarrow{A}_i$. Remember that the valency of ∂A_i is defined as the valency of $-\partial \overrightarrow{A}_i$.

(i) m_i^ν was defined to be the smallest positive integer such that

$$f^{m_i^\nu}(\partial_\nu \overrightarrow{A}_i) = \partial_\nu \overrightarrow{A}_i, \quad \nu = 0, 1.$$

Since A_i is not amphidrome,

$$f^m(\partial_\nu \overrightarrow{A}_i) = \partial_\nu \overrightarrow{A}_i$$

if and only if

$$f^m(A_i) = A_i, \quad \nu = 0, 1.$$

This proves $m_i^0 = m_i^1$.

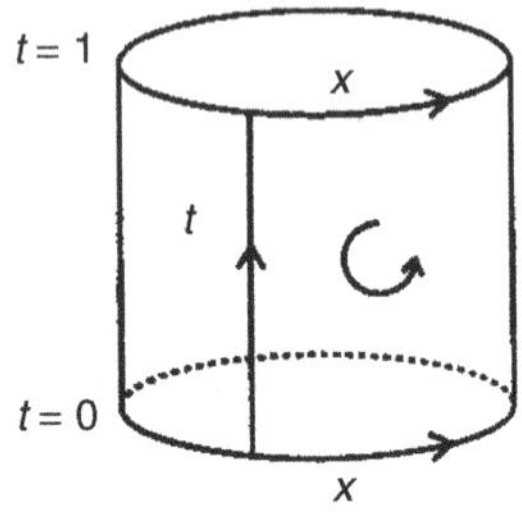

Fig. 2.2 The orientation of the annulus used in Corollary 2.1

(ii) Let us give an orientation to $[0, 1] \times S^1$ as Fig. 2.2 indicates.
Then the orientation of $\{1\} \times S^1$ given by the x-direction is

$$-\partial_1(\overrightarrow{[0, 1] \times S^1})$$

and the orientation of $\{0\} \times S^1$ given by the x-direction is

$$\partial_0(\overrightarrow{[0, 1] \times S^1}).$$

Let m denote the common number $m_i^0 = m_i^1$, and consider

$$f^m : A_i \to A_i$$

as the homeomorphism (not yet normalized) $f : A \to A$ of Lemma 2.1.

Take a parametrization

$$\phi : [0, 1] \times S^1 \to A_i$$

for which

$$f^m \phi(0, x) = \phi(0, x + a)$$

and

$$f^m \phi(1, x) = \phi(1, x + b)$$

for some $a, b \in \mathbf{Q}$, and identify A_i with $[0, 1] \times S^1$ through ϕ.

Then by the geometric meaning of $\delta_i^\nu / \lambda_i^\nu$ ($\nu = 0, 1$), we have

$$\frac{\delta_i^1}{\lambda_i^1} \equiv b \pmod 1,$$

$$\frac{\delta_i^0}{\lambda_i^0} \equiv -a \pmod 1.$$

Recall that if, as in Lemma 2.1,

$$f^m([0, 1] \times \{0\})$$

is lifted to a curve in $[0, 1] \times \mathbf{R}$ joining $(0, a)$ to $(1, b + n)$, then

$$s(A_i) = -(b + n - a).$$

(Remember the Convention (†) on the sign of a Dehn twist.)
Therefore,

$$s(A_i) + \frac{\delta_i^0}{\lambda_i^0} + \frac{\delta_i^1}{\lambda_i^1} \equiv -(b + n - a) - a + b \equiv 0 \pmod 1$$

□

Corollary 2.2. *Let*

$$f : A \to A$$

be a linear twist with respect to a parametrization

$$\phi : [0, 1] \times S^1 \to A.$$

Then the equation giving the linear twist is determined, up to the ambiguity of an integer l, by the screw number $s(A)$ and the valency

$$(m^0, \lambda^0, \sigma^0)$$

of

$$\partial_0 A(\{0\} \times S^1)$$

as follows:

$$f\phi(t, x) = \phi\left(t, x - s(A)t - \frac{\delta^0}{\lambda^0} + l\right), \quad l \in \mathbf{Z},$$

where δ^0 is determined by

$$\delta^0 \sigma^0 \equiv 1 \pmod{\lambda^0}, \quad 0 \leq \delta^0 < \lambda^0.$$

The proof is immediate from the argument of Corollary 2.1.

Lemma 2.2 (Uniqueness of linearization). *Let*

$$f, f' : A \to A$$

be linear twists of an annulus A. Suppose that

$$f \mid \partial A = f' \mid \partial A,$$

and that the screw number of f in A is equal to the screw number of f' in A. Then there is an isotopy

$$h_\tau : A \to A, \quad 0 \le \tau \le 1,$$

such that

(i) $h_0 = id_A$,
(ii) $h_\tau^{-1}(f' \mid \partial A) h_\tau = f' \mid \partial A (= f \mid \partial A)$ *on* ∂A, *and*
(iii) $f = h_1^{-1} f' h_1$.

Remark 2.2. This shows that the quality of being linear is independent of the parametrization ϕ up to a sort of isotopy given by conditions (i), (ii) and (iii) above. Essentially this isotopy is boundary equivariant.

Proof (of Lemma 2.2). Let ϕ and $\phi' : [0,1] \times S^1 \to A$ be the parametrizations for which f and f' are linear twists respectively. By considering $\phi'\rho$ instead of ϕ', if necessary, we may assume

$$\phi(\{0\} \times S^1) = \phi'(\{0\} \times S^1)$$

and

$$\phi(\{1\} \times S^1) = \phi'(\{1\} \times S^1),$$

where

$$\rho : [0,1] \times S^1 \to [0,1] \times S^1$$

is defined by

$$\rho(t, x) = (1 - t, -x).$$

For a while we confine ourselves to the boundary component

$$\partial_0 A = \phi(\{0\} \times S^1),$$

and denote the restrictions

$$\phi \mid \{0\} \times S^1 : \{0\} \times S^1 \to \partial_0 A$$

and

$$\phi' \mid \{0\} \times S^1 : \{0\} \times S^1 \to \partial_0 A$$

simply by

$$\phi : S^1 \to \partial_0 A$$

and

$$\phi' : S^1 \to \partial_0 A,$$

respectively. By the definition of a (**Q**-) linear twist, the action of

$$f \mid \partial_0 A = f' \mid \partial_0 A$$

on $\partial_0 A$ is topologically equivalent to a rotation of finite order, say, $\lambda > 0$.

Since

$$\phi : S^1 \to \partial_0 A$$

and

$$\phi' : S^1 \to \partial_0 A$$

are "linear" parametrizations for this same action we have

$$(\phi')^{-1}\phi\left(x + \frac{1}{\lambda}\right) = (\phi')^{-1}\phi(x) + \frac{1}{\lambda}$$

(Recall $S^1 = \mathbf{R}/\mathbf{Z}$ here). However,

$$(\phi')^{-1}\phi : S^1 \to S^1$$

need not be a "linear" rotation. This requires additional technicality in the first half of the proof below.

Working in the universal covering $\mathbf{R}$, we define an isotopy

$$l_\tau : \mathbf{R} \to \mathbf{R}, \quad 0 \leq \tau \leq 1,$$

by

$$l_\tau(x) = (1-\tau)(\phi')^{-1}\phi(x) + \tau x, \quad x \in \mathbf{R}$$

Obviously we have

$$l_0 = (\phi')^{-1}\phi, l_1 = id_{\mathbf{R}},$$

and

$$l_\tau(x + 1/\lambda) = l_\tau(x) + 1/\lambda.$$

The last property assures that l_τ projects to an isotopy of $S^1 = \mathbf{R}/\mathbf{Z}$ which we denote by the same notation $l_\tau : S^1 \to S^1$. It satisfies

$$l_0 = (\phi')^{-1}\phi, l_1 = id_{S^1},$$

and

$$l_\tau(x + 1/\lambda) = l_\tau(x) + 1/\lambda.$$

Define

$$g_\tau : \partial_0 A \to \partial_0 A, (0 \leq \tau \leq 1)$$

by

$$g_\tau = \phi' l_\tau \phi^{-1}.$$

Then

$$g_0 = id$$

and

$$g_1 = \phi'\phi^{-1}.$$

Also g_τ satisfies the condition

$$g_\tau^{-1}(f' \mid \partial_0 A) g_\tau = f' \mid \partial_0 A \text{ on } \partial_0 A.$$

To see this, recall that

$$(\phi')^{-1}(f' \mid \partial_0 A)\phi' : S^1 \to S^1$$

and

$$\phi^{-1}(f' \mid \partial_0 A)\phi = \phi^{-1}(f \mid \partial_0 A)\phi : S^1 \to S^1$$

are the same rotation of order λ;

$$\phi^{-1}(f' \mid \partial_0 A)\phi(x) = (\phi')^{-1}(f' \mid \partial_0 A)\phi'(x) = x + \frac{\delta}{\lambda}, \tag{2.1}$$

where δ is an integer such that $\gcd(\delta, \lambda) = 1$. On the other hand,

$$\begin{aligned}\phi^{-1} g_\tau^{-1}(f' \mid \partial_o A) g_\tau \phi(x) &= l_\tau^{-1}(\phi')^{-1}(f' \mid \partial_0 A)\phi' l_\tau(x) \\ &= l_\tau^{-1}\left(l_\tau(x) + \frac{\delta}{\lambda}\right) \quad \text{by} \quad (2.1) \\ &= x + \frac{\delta}{\lambda}.\end{aligned}$$

Thus

$$\phi^{-1} g_\tau^{-1}(f' \mid \partial_0 A) g_\tau \phi = \phi^{-1}(f' \mid \partial_0 A)\phi,$$

so

$$g_\tau^{-1}(f' \mid \partial_0 A) g_\tau = f' \mid \partial_0 A$$

as asserted.

We extend $g_\tau : \partial_0 A \to \partial_0 A$ to an isotopy

$$\overline{g}_\tau : A \to A, \quad 0 \le \tau \le 1,$$

in such a way that

$$\overline{g}_0 = id_A, \quad \overline{g}_\tau \mid \partial_0 A = g_\tau, \quad \text{and} \quad \overline{g}_\tau \mid \partial_1 A = \text{identity}.$$

Then the isotopy $\overline{g}_\tau$ satisfies conditions (i) and (ii) of Lemma 2.2, and

$$(\overline{g}_1)^{-1} f' \overline{g}_1 : A \to A$$

is a linear twist with respect to the parametrization

$$(\overline{g}_1)^{-1}\phi' : [0,1] \times S^1 \to A.$$

Moreover, this parametrization $(\overline{g}_1)^{-1}\phi'$ satisfies

$$(\overline{g}_1)^{-1}\phi' \mid \{0\} \times S^1 = \phi \mid \{0\} \times S^1$$

Therefore, taking

$$(\overline{g}_1)^{-1} f' \overline{g}_1$$

instead of f' if necessary, we may assume the parametrization

$$\phi' : [0, 1] \times S^1 \to A$$

for f' satisfies

$$\phi' \mid \{0\} \times S^1 = \phi \mid \{0\} \times S^1.$$

Applying the same argument to $\{1\} \times S^1$, we may assume

$$\phi' \mid \{0, 1\} \times S^1 = \phi \mid \{0, 1\} \times S^1.$$

This completes the preparatory argument.

Now by the technique of the proof of Lemma 2.1, there exists an isotopy

$$\Phi_\tau : [0, 1] \times S^1 \to [0, 1] \times S^1, \quad 0 \leq \tau \leq 1,$$

such that

1. $\Phi_0 = (\phi')^{-1}\phi$,

2. Φ_1 is linear, that is
$$\Phi_1(t, x) = (t, x + at + b)$$
for some $a, b \in \mathbf{Q}$, and

3. the restriction
$$\Phi_\tau \mid \{0, 1\} \times S^1$$
equals $(\phi')^{-1}\phi \mid \{0, 1\} \times S^1$ (= the identity of $\{0, 1\} \times S^1$).

Define

$$h_\tau : A \to A, \quad 0 \leq \tau \leq 1,$$

by

$$h_\tau = \phi' \Phi_\tau \phi^{-1}.$$

Then h_τ satisfies (i) $h_0 = id_A$, (ii) $h_\tau \mid \partial A = id_{\partial A}$, in particular

$$h_\tau^{-1}(f' \mid \partial A)h_\tau = f' \mid \partial A.$$

Moreover, one can verify

$$(h_1^{-1} f' h_1)\phi(t, x) = \phi(t, x + ct + d) \text{ for some } c, d \in \mathbf{Q}.$$

Thus the homeomorphisms f and $h_1^{-1} f' h_1 : A \to A$ are both linear with respect to the same ϕ; they coincide on the boundary ∂A, and have the same screw number by the assumption. Therefore, (iii) $f = h_1^{-1} f' h_1$. □

Putting together Lemmmas 2.1 and 2.2, and generalizing them to a disjoint union of annuli, we obtain the following theorem.

Theorem 2.3. (i) (LINEARIZATION) *Let*

$$\mathscr{A} = \bigcup_{i=1}^{r} A_i$$

be a disjoint union of annuli,

$$f : \mathscr{A} \to \mathscr{A}$$

a homeomorphism such that

$$f(A_i) = A_{i+1}, \quad i = 1, 2, \ldots, r-1,$$

and $f(A_r) = A_1$. *Suppose that*

$$f \mid \partial\mathscr{A} : \partial\mathscr{A} \to \partial\mathscr{A}$$

is periodic and that no annulus in $\mathscr{A}$ *is amphidrome with respect to* f. *Then* f *is rel.*∂ *isotopic to a homeomorphism* $f' : \mathscr{A} \to \mathscr{A}$ *such that*

$$(f')^r \mid A_i : A_i \to A_i$$

is a linear twist for each $i = 1, 2, \ldots, r$.
(ii) (UNIQUENESS OF LINEARIZATION) *Let*

$$f, f' : \mathscr{A} \to \mathscr{A}$$

be homeomorphisms such that

$$f(A_i) = f'(A_i) = A_{i+1}, \quad i = 1, 2, \ldots, r-1,$$

and

$$f(A_r) = f'(A_r) = A_1.$$

Suppose that $(f)^r \mid A_i$ *and*

$$(f')^r \mid A_i : A_i \to A_i$$

are linear twists for each $i = 1, 2, \ldots, r$, *and that* $f \mid \partial\mathscr{A} = f' \mid \partial\mathscr{A}$, *and that*

$$f, f' : \mathscr{A} \to \mathscr{A}$$

are mutually isotopic by a rel.∂ isotopy. Then there is an isotopy

$$h_\tau : \mathscr{A} \to \mathscr{A}, \quad 0 \le \tau \le 1,$$

such that

1. $h_0 = id_{\mathscr{A}}$
2. $h_\tau^{-1}(f' \mid \partial\mathscr{A})\, h_\tau = f' \mid \partial\mathscr{A} (= f \mid \partial\mathscr{A})$ *on* $\partial\mathscr{A}$, *and*
3. $f = h_1^{-1} f' h_1$.

Proof (of (i) LINEARIZATION*).* Let

$$\phi : [0,1] \times S^1 \to A_1$$

be a parametrization for which

$$f^r : A_1 \to A_1$$

is "linear" on ∂A_1. We adopt

$$f^{i-1}\phi : [0,1] \times S^1 \to A_i$$

as a parametrization of A_i, $i = 1, 2, \dots, r$. By essentially the same argument as in Lemma 2.1,

$$f \mid A_r : A_r \to A_1$$

is *rel.∂* isotopic to a map

$$f' \mid A_r : A_r \to A_1$$

which is "linear" with respect to the parametrizations

$$f^{r-1}\phi : [0,1] \times S^1 \to A_r$$

and

$$\phi : [0,1] \times S^1 \to A_1.$$

We define

$$f' : \mathscr{A} \to \mathscr{A}$$

by setting

$$f' \mid A_i = f \mid A_i : A_i \to A_{i+1},$$

for $i = 1, 2, \dots, r-1$, and by taking the above $f' \mid A_r : A_r \to A_1$, for $i = r$.

Then

$$(f')^r \mid A_1 : A_1 \to A_1$$

satisfies

$$\begin{aligned}(f')^r\phi(t,x) &= (f' \mid A_r)(f' \mid A_{r-1})\cdots(f' \mid A_1)\phi(t,x)\\ &= (f' \mid A_r)f^{r-1}\phi(t,x)\\ &= \phi(t, x+at+b), \text{ for some } a, b \in \mathbf{Q}\end{aligned}$$

Thus

$$(f')^r \mid A_1 : A_1 \to A_1$$

is a linear twist with respect to $\phi : [0,1] \times S^1 \to A_1$.

Similarly, if $r \geq 2$,

$$(f')^r \mid A_2 : A_2 \to A_2$$

satisfies

$$\begin{aligned}(f')^r f\phi(t,x) &= (f' \mid A_1)(f' \mid A_r)\cdots(f' \mid A_2)(f' \mid A_1)\phi(t,x)\\ &= (f \mid A_1)(f' \mid A_r)f^{r-1}\phi(t,x)\\ &= (f \mid A_1)\phi(t, x+at+b)\\ &= f\phi(t, x+at+b).\end{aligned}$$

Thus

$$(f')^r \mid A_2 : A_2 \to A_2$$

is a linear twist with respect to

$$f\phi : [0,1] \times S^1 \to A_2,$$

and so on. This proves (i). □

Proof (of (ii) UNIQUENESS OF LINEARIZATION*)*. Applying Lemma 2.2 to

$$f^r \mid A_1 : A_1 \to A_1$$

and

$$(f')^r \mid A_1 : A_1 \to A_1,$$

we find an isotopy

$$g_\tau^{(1)} : A_1 \to A_1, \quad 0 \leq \tau \leq 1,$$

such that

$$g_0^{(1)} = id_{A_1}, (g_\tau^{(1)})^{-1}((f')^r \mid \partial A_1)g_\tau^{(1)} = (f')^r \mid \partial A_1,$$

and

$$f^r \mid A_1 = (g_1^{(1)})^{-1}((f')^r \mid A_1)g_\tau^{(1)}.$$

Define isotopies

$$g_\tau^{(i)} : A_i \to A_i, \quad 0 \leq \tau \leq 1, \quad i = 2, \ldots, r$$

by the formula

$$g_\tau^{(i)} = (f \mid A_{i-1}) \cdots (f \mid A_1) g_\tau^{(1)} (f \mid A_1)^{-1} \cdots (f \mid A_{i-1})^{-1}$$

(or equivalently, by an inductive formula

$$g_\tau^{(i)} = (f \mid A_{i-1}) g_\tau^{(i-1)} (f \mid A_{i-1})^{-1},$$

$i = 2, \ldots, r$) and define an isotopy

$$g_\tau : \mathscr{A} \to \mathscr{A}, \quad 0 \leq \tau \leq 1,$$

by setting

$$g_\tau \mid A_i = g_\tau^{(i)}, \quad i = 1, 2, \ldots, r$$

Then g_τ satisfies $g_o = id_{\mathscr{A}}$,

$$g_\tau^{-1}(f' \mid \partial\mathscr{A}) g_\tau = f' \mid \partial\mathscr{A},$$

and

$$(g_1^{-1} f' g_1)^r \mid A_1 = f^r \mid A_1.$$

Therefore, replacing f' by $g_1^{-1} f' g_1$ if necessary, we may assume

$$(f')^r \mid A_1 = f^r \mid A_1.$$

By the assumption,

$$f : (\mathscr{A}, \partial\mathscr{A}) \to (\mathscr{A}, \partial\mathscr{A})$$

is rel.∂ isotopic to

$$f' : (\mathscr{A}, \partial\mathscr{A}) \to (\mathscr{A}, \partial\mathscr{A}).$$

In particular, there is an isotopy

$$h_\tau^{(2)} : A_2 \to A_2, \quad 0 \leq \tau \leq 1,$$

such that $h_0^{(2)} = id_{A_2}$, $h_\tau^{(2)} \mid \partial A_2 = id_{\partial A_2}$, and $h_1^{(2)}(f \mid A_1) = f' \mid A_1$.

Then

$$(f \mid A_2)(h_1^{(2)})^{-1} : A_2 \to A_3$$

is rel.∂ isotopic to

$$f' \mid A_2 : A_2 \to A_3,$$

so there is an isotopy

$$h_\tau^{(3)} : A_3 \to A_3, \quad 0 \leq \tau \leq 1,$$

such that $h_0^{(3)} = id_{A_3}$, $h_\tau^{(3)} \mid \partial A_3 = id_{\partial A_3}$, and $h_1^{(3)}(f \mid A_2)(h_1^{(2)})^{-1} = f' \mid A_2$.

Proceeding in this way, we can construct isotopies

$$h_\tau^{(i)} : A_i \to A_i, \quad 0 \le \tau \le 1, i = 2, \dots, r,$$

such that $h_0^{(i)} = id_{A_i}, h_\tau^{(i)} \mid \partial A_i = id_{\partial A_i}$, and

$$h_1^{(i)}(f \mid A_{i-1})(h_1^{(i-1)})^{-1} = f' \mid A_{i-1}.$$

We will examine

$$(f \mid A_r)(h_1^{(r)})^{-1} : A_r \to A_1.$$

For this purpose, set

$$P = h_1^{(r)}(f \mid A_{r-1})(f \mid A_{r-2}) \cdots (f \mid A_1) : A_1 \to A_r.$$

Then

$$\begin{aligned} P &= h_1^{(r)}(f \mid A_{r-1})(h_1^{(r-1)})^{-1} h_1^{(r-1)}(f \mid A_{r-2})(h_1^{(r-2)})^{-1} \cdots \\ &\quad h_1^{(3)}(f \mid A_2)(h_1^{(2)})^{-1} h_1^{(2)}(f \mid A_1) \\ &= (f')^{r-1} \mid A_1. \end{aligned}$$

We have

$$(f \mid A_r)(h_1^{(r)})^{-1} P = f^r \mid A_1 = (f')^r \mid A_1 = (f' \mid A_r) P$$

implying

$$(f \mid A_r)(h_1^{(r)})^{-1} = f' \mid A_r : A_r \to A_1$$

Finally define an isotopy

$$h_\tau : \mathscr{A} \to \mathscr{A}, \quad 0 \le \tau \le 1,$$

by setting

$$\begin{aligned} h_\tau \mid A_1 &= id, \\ h_\tau \mid A_i &= h_\tau^{(i)}, \quad i = 2, \dots, r. \end{aligned}$$

Then,

$$h_0 = id_{\mathscr{A}}, h_\tau \mid \partial\mathscr{A} = id_{\partial\mathscr{A}}, \text{ and } f' = h_1 f h_1^{-1} \text{ (i.e., } f = h_1^{-1} f' h_1)$$

as asserted. □

2.4 Amphidrome Annuli

Lemma 2.3 (SPECIALIZATION). *Let A be an annulus,*

$$f : A \to A$$

an (orientation-preserving) homeomorphism which interchanges the boundary components. Suppose

$$f \mid \partial A : \partial A \to \partial A$$

is periodic. Then there exists a rel.∂ isotopy

$$f_\tau : A \to A, \quad 0 \leq \tau \leq 1,$$

such that $f_0 = f$ and

$$f_1 : A \to A$$

is a special twist.

We need a sublemma.

Sublemma 1 Let *A and*

$$f : A \to A$$

be as in Lemma 2.3. Then there is a parametrization

$$\phi : [0,1] \times S^1 \to A$$

such that

$$f\phi(0,x) = \phi(1, -x + a)$$
$$f\phi(1,x) = \phi(0, -x - a)$$

for some $a \in Q$.

Proof. Let $\partial A = \partial_0 A \cup \partial_1 A$. Since

$$f^2 \mid \partial_0 A : \partial_0 A \to \partial_0 A$$

is a rotation of finite order, there is a parametrization

$$\varphi : S^1 \to \partial_0 A$$

such that

$$f^2 \varphi(x) = \varphi(x - 2b)$$

for some $b \in \mathbf{Q}$. Take a parametrization

$$\psi : S^1 \to \partial_1 A$$

for which

$$f\varphi(x) = \psi(-x + a)$$

holds, where $a \in \mathbf{Q}$ is arbitrary at this point.

Now,

$$\begin{aligned} f\psi(x) &= (f^2\varphi)(-x + a) \\ &= \varphi(-x + a - 2b) \end{aligned}$$

We have already chosen the number a satisfying

$$f\varphi(x) = \psi(-x + a)$$

and we want a to satisfy also

$$f\psi(x) = \varphi(-x - a).$$

To attain this $a \in \mathbf{Q}$ must be chosen so that

$$a - 2b \equiv -a \pmod 1,$$

that is

$$a \equiv b \text{ or } b + \frac{1}{2} \pmod 1.$$

This ambiguity cannot be settled now.

Since

$$\varphi : S^1 \to \partial_0 A$$

and

$$\psi : S^1 \to \partial_1 A$$

are homotopic as maps from S^1 into A, there is a homeomorphism

$$\phi : [0, 1] \times S^1 \to A$$

such that $\phi(0, x) = \varphi(x)$ and $\phi(1, x) = \psi(x)$.

Then

$$\begin{aligned} f\phi(0, x) &= \phi(1, -x + a), \\ f\phi(1, x) &= \phi(0, -x - a), \end{aligned}$$

as required. But remember that there are two possible values for $a \pmod 1$. □

This sublemma has the following corollary.

Corollary 2.3. *Let*

$$\mathscr{A} = \bigcup_{i=1}^{r} A_i$$

be a disjoint union of annuli,

$$f : \mathscr{A} \to \mathscr{A}$$

a homeomorphism such that

$$f \mid \partial\mathscr{A} : \partial\mathscr{A} \to \partial\mathscr{A}$$

is periodic. Suppose A_i be an amphidrome annulus in $\mathscr{A}$ with respect to f. Let

$$(m_i^0, \lambda_i^0, \sigma_i^0)$$

and

$$(m_i^1, \lambda_i^1, \sigma_i^1)$$

be the valencies of $\partial_0 A$ and $\partial_1 A$, respectively, and $s(A_i)$ the screw number of f in A_i. Then

(i) $m_i^0 = m_i^1 =$ *an even number,*
(ii) $(\lambda_i^0, \sigma_i^0) = (\lambda_i^1, \sigma_i^1)$
(iii) $(1/2)s(A_i) + \delta_i/\lambda_i$ *is an integer, where δ_i is an integer determined by $\sigma_i \delta_i \equiv 1 \pmod{\lambda_i}$ and $0 \leq \delta_i < \lambda_i$. (Here λ_i denotes $\lambda_i^0 = \lambda_i^1$, and σ_i denotes $\sigma_i^0 = \sigma_i^1$.)*
(iv) (λ_i, σ_i)*is uniquely determined by $s(A_i)$.*

Proof. Let k be the smallest positive integer such that $f^k(A_i) = A_i$. Since A_i is amphidrome, f^k interchanges the boundary components of A_i. Thus $2k$ is the smallest positive integer such that $f^{2k}(A_i) = A_i$ does not interchange the boundary components. This implies (i) $m_i^0 = m_i^1 = 2k$.

Obviously,

$$f^k \mid \partial_0 A : \partial_0 A \to \partial_1 A$$

is equivariant with respect to the actions of $f^{2k} \mid \partial_0 A$ on $\partial_0 A$ and $f^{2k} \mid \partial_1 A$ on $\partial_1 A$. This proves (ii) $(\lambda_i^0, \sigma_i^0) = (\lambda_i^1, \sigma_i^1)$.

To prove (iii), consider

$$f^k : A_i \to A_i$$

as f in sublemma 1 and take a parametrization

$$\phi : [0, 1] \times S^1 \to A_i$$

there. We identify A_i with $[0, 1] \times S^1$ through ϕ. We give to $[0, 1] \times S^1$ the same orientation as in the proof of Corollary 2.1 (Fig. 2.2).

Note that

$$f^k(0, x) = (1, -x + a), f^k(1, x) = (0, -x - a).$$

Let us pass to the universal covering $[0, 1] \times \mathbf{R}$.

Let $\tilde{I}$ be the lift of the arc

$$f^k([0, 1] \times \{0\})$$

which starts at

$$(0, -a) \in [0, 1] \times \mathbf{R}.$$

Then it ends at $(1, a + n)$ for some $n \in \mathbf{Z}$. Thus there is a lift

$$\tilde{f^k} : [0, 1] \times \mathbf{R} \to [0, 1] \times \mathbf{R}$$

of

$$f^k : [0, 1] \times S^1 \to [0, 1] \times S^1$$

satisfying

$$\tilde{f^k}(0, x) = (1, -x + a + n),$$
$$\tilde{f^k}(1, x) = (0, -x - a).$$

We have

$$(\tilde{f^k})^2(0, x) = (0, x - 2a - n),$$
$$(\tilde{f^k})^2(1, x) = (1, x + 2a + n).$$

The curve $f^{2k}([0, 1] \times \{0\})$ is lifted to a curve joining $(0, -2a-n)$ to $(1, 2a+n)$. This implies

$$s(A_i) = -4a - 2n. \tag{2.2}$$

(Recall the convention of the sign of a Dehn twist in Convention (†).)

On the other hand, by the geometric meaning of δ_i/λ_i, we have

$$\frac{\delta_i}{\lambda_i} \equiv 2a \pmod 1.$$

Thus

$$\frac{1}{2}s(A_i) + \frac{\delta_i}{\lambda_i} \equiv -2a - n + 2a \equiv 0 \pmod 1,$$

which proves (iii).

Since $0 \leq \sigma_i < 1$ and $\sigma_i \delta_i \equiv 1 \pmod{\lambda_i}$, assertion (iv) follows from (iii). □

Now we can fix the ambiguity of the number a unsettled in Sublemma 1. We observed in the above proof that

$$\frac{1}{2}s(A_i) = -2a - n.$$

Hence by shifting a by $1/2$ if necessary, we can make n an even number. In other words, we can (and will) take a so that

$$a \equiv -\frac{1}{4}s(A_i) \pmod 1.$$

Let us restate Sublemma 1 as Corollary 2.4, taking this choice into account.

Corollary 2.4. *Let A and*

$$f : A \to A$$

be as in Lemma 2.3. *Let $s(A)$ be the screw number of f in A. Then there is a parametrization*

$$\phi : [0, 1] \times S^1 \to A$$

such that

$$f\phi(0, x) = \phi\left(1, -x - \frac{1}{4}s(A)\right),$$

$$f\phi(1, x) = \phi\left(0, -x + \frac{1}{4}s(A)\right).$$

Remark 2.3. In this way f^2 rotates both sides of A by half the screw number in *opposite* directions.

Proof (of Lemma 2.3). Let

$$\phi : [0, 1] \times S^1 \to A$$

be the parametrization of Corollary 2.4, and identify A with $[0, 1] \times S^1$ through this ϕ. Let

$$\tilde{f} : [0, 1] \times \mathbf{R} \to [0, 1] \times \mathbf{R}$$

be the lift of

$$f : [0, 1] \times S^1 \to [0, 1] \times S^1$$

such that

$$\tilde{f}(1, x) = \left(0, -x + \frac{1}{4}s(A)\right).$$

Then

$$\tilde{f}(0, x) = (1, -x - \frac{1}{4}s(A) + n)$$

for some $n \in \mathbf{Z}$, but this n must be 0 because, as we observed in the proof of Corollary 2.3, if the curve

$$\tilde{f}([0,1] \times \{0\})$$

connects $(0, 1/4s(A))$ and $(1, -1/4s(A) + n)$, then

$$s(A) = -4(-\frac{1}{4}s(A)) - 2n.$$

(See (2.2) in the proof.) This implies $n = 0$, and we get

$$\tilde{f}(0, x) = \left(1, -x - \frac{1}{4}s(A)\right).$$

Define a piecewise-linear homeomorphism

$$\tilde{f}' : [0,1] \times \mathbf{R} \to [0,1] \times \mathbf{R}$$

by setting

$$\tilde{f}'(t, x) = \begin{cases} \left(1 - t, -x + \frac{3}{4}s(A)\left(t - \frac{1}{3}\right)\right), & 0 \le t \le \frac{1}{3}, \\ (1 - t, -x), & \frac{1}{3} \le t \le \frac{2}{3}, \\ \left(1 - t, -x + \frac{3}{4}s(A)\left(t - \frac{2}{3}\right)\right), & \frac{2}{3} \le t \le 1. \end{cases}$$

Note that

$$\tilde{f}' \mid \{0,1\} \times \mathbf{R} = \tilde{f} \mid \{0,1\} \times \mathbf{R}.$$

This homeomorphism $\tilde{f}'$ projects to a special twist

$$f' : [0,1] \times S^1 \to [0,1] \times S^1.$$

By essentially the same argument in the proof of Lemma 2.1,

$$f : [0,1] \times S^1 \to [0,1] \times S^1$$

is *rel.*∂ isotopic to f'. This completes the proof of Lemma 2.3. □

The following corollary will be obvious from the above argument.

Corollary 2.5. *Let $f : A \to A$ be a special twist with respect to a parametrization*

$$\phi : [0,1] \times S^1 \to A.$$

Then, in contradistinction with the case of linear twists (Corollary 2.2), the equation defining a special twist is uniquely determined by the screw number $s(A)$ as follows:

$$f\phi(t,x) = \begin{cases} \phi\left(1-t,\ -x+\frac{3}{4}s(A)\left(t-\frac{1}{3}\right)\right), & 0 \le t \le \frac{1}{3}, \\ \phi(1-t,\ -x), & \frac{1}{3} \le t \le \frac{2}{3}, \\ \phi\left(1-t,\ -x+\frac{3}{4}s(A)\left(t-\frac{2}{3}\right)\right), & \frac{2}{3} \le t \le 1. \end{cases}$$

Lemma 2.4 (Uniqueness of specialization). *Let*

$$f, f' : A \to A$$

be special twists of an annulus A. Suppose that

$$f \mid \partial A = f' \mid \partial A$$

and that the screw number of f in A is equal to the screw number of f' in A. Then there is an isotopy

$$h_\tau : A \to A, \quad 0 \le \tau \le 1,$$

such that

(i) $h_0 = id_A$,
(ii) $h_\tau^{-1}(f' \mid \partial A)h_\tau = f' \mid \partial A\ (= f \mid \partial A) on \partial A$, *and*
(iii) $f = h_1^{-1} f' h_1$.

Proof (of Lemma 2.4). The idea is the same as in Lemma 2.2. Let ϕ and

$$\phi' : [0.1] \times S^1 \to A$$

be the parametrizations with respect to which f and f' are special twists. After a preliminary isotopy, we may assume

$$\phi \mid \{0,1\} \times S^1 = \phi' \mid \{0,1\} \times S^1.$$

(Cf. the proof of Lemma 2.2.). Then, the next Claim shows that ϕ and ϕ' differ by an *even* number of full twists.

Claim (A). There exist lifts

$$\tilde{\phi}, \tilde{\phi}' : [0,1] \times \mathbf{R} \to \tilde{A}$$

of

$$\phi, \phi' : [0,1] \times S^1 \to A$$

such that

$$(\tilde{\phi}')^{-1}\tilde{\phi}(0,x) = (0, x+m),$$
$$(\tilde{\phi}')^{-1}\tilde{\phi}(1,x) = (1, x-m),$$

for some $m \in \mathbf{Z}$.

Proof (of Claim (A)). For simplicity, we identify A with $[0,1] \times S^1$ through ϕ'. By Corollary 2.5,

$$f' : [0,1] \times S^1 \to [0,1] \times S^1$$

is lifted to

$$\tilde{f}' : [0,1] \times \mathbf{R} \to [0,1] \times \mathbf{R}$$

such that the image of the interval $[0,1] \times \{0\}$ under $\widetilde{f}'$ is a piecewise-linear arc connecting $(0, 1/4s)$ and $(1, -1/4s)$, where $s = s(A)$. Since $f \mid \partial A = f' \mid \partial A$, and the screw number $s(A)$ is common to f and f',

$$f : [0,1] \times S^1 \to [0,1] \times S^1$$

can also be lifted to

$$\tilde{f} : [0,1] \times \mathbf{R} \to [0,1] \times \mathbf{R}$$

such that $f([0,1] \times \{0\})$ is a (not necessarily piecewise-linear) arc connecting $(0, 1/4s)$ and $(1, -1/4s)$. In particular,

$$\tilde{f}(0,x) = \left(1, -x - \frac{1}{4}s\right), \quad \tilde{f}(1,x) = \left(0, -x + \frac{1}{4}s\right).$$

By Corollary 2.5, there is a lift

$$\tilde{\phi} : [0,1] \times \mathbf{R} \to \tilde{A} (= [0,1] \times \mathbf{R} \text{ via } \tilde{\phi}')$$

such that

$$(\tilde{\phi})^{-1} \tilde{f} \tilde{\phi}([0,1] \times \{0\})$$

is a piecewise-linear arc connecting $(0, 1/4s)$ and $(1, -1/4s)$. Since

$$\phi \mid \{0,1\} \times S^1 = \phi' \mid \{0,1\} \times S^1,$$

$\tilde{\phi}$ satisfies

$$\tilde{\phi}(0,x) = (0, x+m),$$
$$\tilde{\phi}(1,x) = (1, x+n),$$

for some $m, n \in \mathbf{Z}$. Then

$$\tilde{\phi}^{-1} \tilde{f} \tilde{\phi}(0,x) = \left(1, -x - m - n - \frac{1}{4}s\right),$$
$$\tilde{\phi}^{-1} \tilde{f} \tilde{\phi}(1,x) = \left(0, -x - m - n + \frac{1}{4}s\right).$$

By our choice of $\tilde{\phi}$, $m + n = 0$. Thus

$$(\tilde{\phi}')^{-1}\tilde{\phi} : [0, 1] \times \mathbf{R} \to [0, 1] \times \mathbf{R}$$

satisfies the formula stated in Claim (A). □

Proof (of Lemma 2.4: continued). Just as in the proof of Lemma 2.2, there is an isotopy

$$\Phi_\tau : [0, 1] \times S^1 \to [0, 1] \times S^1$$

such that

1. $\Phi_0 = (\phi')^{-1}\phi$,
2. Φ_1 is linear;

$$\Phi_1(t, x) = (t, x - 2mt + m),$$

   m being the integer of Claim (A).
3. The restriction

$$\Phi_\tau \mid \{0, 1\} \times S^1$$

   equals

$$(\phi')^{-1}\phi \mid \{0, 1\} \times S^1$$

   which is the identity on $\{0, 1\} \times S^1$.

Claim (B). If $f' : A \to A$ is a special twist with respect to a parametrization

$$\phi' : [0, 1] \times S^1 \to A,$$

then f' is also a special twist with respect to

$$\phi'\Phi_1 : [0, 1] \times S^1 \to A,$$

where

$$\Phi_1 : [0, 1] \times S^1 \to [0, 1] \times S^1$$

is given by

$$\Phi_1(t, x) = (t, x - 2mt + m)$$

for some $m \in \mathbf{Z}$.

Proof (of Claim (B)). Compute

$$\Phi_1^{-1}(\phi')^{-1} f'\phi'\Phi_1(t, x).$$

But beware that it is here where the full force of Claim (A) is used. □

Define an isotopy

$$h_\tau : A \to A, \quad 0 \leq \tau \leq 1,$$

by

$$h_\tau = \phi' \Phi_\tau \phi^{-1}.$$

Then h_τ satisfies (i) $h_0 = id_A$, (ii) $h_\tau \mid \partial A = id_{\partial A}$. Moreover, using Claim (B), one can verify that

$$h_1^{-1} f' h_1 (= \phi \Phi_1^{-1} (\phi')^{-1} f' \phi' \Phi_1 \phi^{-1})$$

is a special twist with respect to $\phi : [0,1] \times S^1 \to A$. Both $h_1^{-1} f' h_1$ and f are special twists with respect to the same parametrization

$$\phi : [0,1] \times S^1 \to A.$$

Since they coincide on ∂A and have the same screw number in A then (iii) $f = h_1^{-1} f' h_1$. □

Let us generalize Lemmas 2.3 and 2.4 to the case of a disjoint union of annuli.

Theorem 2.4. (i) *(*SPECIALIZATION*) Let*

$$\mathscr{A} = \bigcup_{i=1}^{r} A_i$$

be a disjoint union of annuli,

$$f : \mathscr{A} \to \mathscr{A}$$

a homeomorphism such that

$$f(A_i) = A_{i+1}, \quad i = 1, 2, \ldots, r-1,$$

and $f(A_r) = A_1$. Suppose that

$$f \mid \partial\mathscr{A} : \partial\mathscr{A} \to \partial\mathscr{A}$$

is periodic and that each A_i is amphidrome with respect to f. Then f is rel.∂ isotopic to a homeomorphism

$$f' : \mathscr{A} \to \mathscr{A}$$

such that

$$(f')^r \mid A_i : A_i \to A_i$$

is a special twist for each $i = 1, 2, \ldots, r$.
(ii) *(*UNIQUENESS OF SPECIALIZATION*) Let*

$$f, f' : \mathscr{A} \to \mathscr{A}$$

be homeomorphisms such that

$$f(A_i) = f'(A_i) = A_{i+1}, \quad i = 1, 2, \ldots, r-1,$$

and

$$f(A_r) = f'(A_r) = A_1.$$

Suppose that $f^r \mid A_i$ *and*

$$(f')^r \mid A_i : A_i \to A_i$$

are special twists for each $i = 1, 2, \ldots, r$, *and that*

$$f \mid \partial\mathscr{A} = f' \mid \partial\mathscr{A},$$

and that

$$f, f' : \mathscr{A} \to \mathscr{A}$$

are mutually isotopic by a rel.∂ *isotopy. Then there is an isotopy*

$$h_\tau : A \to A, \quad 0 \leq \tau \leq 1,$$

such that

(i) $h_0 = id_{\mathscr{A}}$,
(ii) $h_\tau^{-1}(f' \mid \partial A)h_\tau = f' \mid \partial A (= f \mid \partial A)$ *on* ∂A, *and*
(iii) $f = h_1^{-1} f' h_1$.

Proof (of (i) SPECIALIZATION*).* Since A_1 is amphidrome,

$$f^r : A_1 \to A_1$$

interchanges the boundary components. Take a parametrization

$$\phi : [0, 1] \times S^1 \to A_1$$

for which

$$f^r : A_1 \to A_1$$

satisfies

$$f^r\phi(0, x) = \phi\left(1, -x - \frac{1}{4}s(A_1)\right),$$

$$f^r\phi(1, x) = \phi\left(0, -x + \frac{1}{4}s(A_1)\right).$$

(See Corollary 2.4).

As in the proof of Theorem 2.3 (i), we adopt

$$f^{i-1}\phi : [0, 1] \times S^1 \to A_i$$

as a parametrization of A_i, $i = 1, 2, \ldots, r$. By our choice of

$$\phi : [0,1] \times S^1 \to A_1,$$

the map

$$f \mid A_r : A_r \to A_1$$

trivially satisfies

$$(f \mid A_r) f^{r-1} \phi(0, x) = \phi\left(1, -x - \frac{1}{4} s(A_1)\right),$$

$$(f \mid A_r) f^{r-1} \phi(1, x) = \phi\left(0, -x + \frac{1}{4} s(A_1)\right).$$

Then

$$f \mid A_r : A_r \to A_1$$

is *rel.*∂ isotopic to a homeomorphism

$$f' \mid A_r : A_r \to A_1$$

defined by

$$(f' \mid A_r) f^{r-1} \phi(t, x) = \begin{cases} \phi\left(1 - t, -x + \frac{3}{4} s(A_1)\left(t - \frac{1}{3}\right)\right), & 0 \le t \le \frac{1}{3}, \\ \phi(1 - t, -x), & \frac{1}{3} \le t \le \frac{2}{3}, \\ \phi\left(1 - t, -x + \frac{3}{4} s(A_1)\left(t - \frac{2}{3}\right)\right), & \frac{2}{3} \le t \le 1. \end{cases}$$

(See the proof of Lemma 2.3). □

We define

$$f' : \mathscr{A} \to \mathscr{A}$$

by setting

$$f' \mid A_i = f \mid A_i : A_i \to A_{i+1},$$

for $i = 1, 2, \ldots, r - 1$, and by taking the above

$$f' \mid A_r : A_r \to A_1$$

for $i = r$. Then by the same argument as in the proof of Theorem 2.3(i), we can show that

$$(f')^r \mid A_i : A_i \to A_i$$

is a special twist with respect to the parametrization

$$f^{i-1} \phi : [0,1] \times S^1 \to A_i,$$

for $i = 1, 2, \ldots, r$. Clearly f' is rel.∂ isotopic to f. We are done.

Proof (of (ii) UNIQUENESS OF SPECIALIZATION*)*. Applying Lemma 2.4 to

$$f^r \mid A_1 : A_1 \to A_1,$$

and

$$(f')^r \mid A_1 : A_1 \to A_1$$

we find an isotopy

$$g_\tau^{(1)} : A_1 \to A_1, \quad 0 \leq \tau \leq 1,$$

such that $g_0^{(1)} = id_{A_1}$,

$$(g_\tau^{(1)})^{-1}((f')^r \mid \partial A_1) g_\tau^{(1)} = (f')^r \mid \partial A_1$$

and

$$f^r \mid A_1 = (g_1^{(1)})^{-1}((f')^r \mid A_1) g_1^{(1)}.$$

Then the rest of the proof is exactly the same as the proof of Theorem 2.3 (ii). This completes the proof. □

Now we are in a position to prove the main theorem of this Chap. 2.

2.5 Proof of Theorem 2.1

Proof (of (i)). We must show that a given pseudo-periodic map

$$f : \Sigma_g \to \Sigma_g$$

is isotopic to a pseudo-periodic map in standard form. Let $\{C_i\}_{i=1}^r$ be the precise system of cut curves subordinate to f (Chap. 1). We may assume

$$f(\mathscr{C}) = \mathscr{C},$$

$\mathscr{C}$ being $\bigcup_{i=1}^r C_i$. We choose a system of annular neighborhoods $\{A_i\}_{i=1}^r$ of $\{C_i\}_{i=1}^r$. We may assume

$$f(\mathscr{A}) = \mathscr{A},$$

where

$$\mathscr{A} = \bigcup_{i=1}^r A_i.$$

By the definition of a pseudo-periodic map,

$$f \mid (\Sigma_g - \mathscr{C}) : \Sigma_g - \mathscr{C} \to \Sigma_g - \mathscr{C}$$

is isotopic to a periodic map. Then

$$f \mid \mathscr{B} : \mathscr{B} \to \mathscr{B}$$

is also isotopic to a periodic map (Cf. [52]), where

$$\mathscr{B} = \Sigma_g - Int(\mathscr{A}),$$

so we may assume that $f \mid \mathscr{B}$ is already periodic.

Decompose the finite set $\{A_i\}_{i=1}^{r}$ into cyclic orbits under the permutation caused by f, and decompose $\mathscr{A}$ into

$$\mathscr{A}^{(1)} \cup \mathscr{A}^{(2)} \cup \cdots \cup \mathscr{A}^{(s)}$$

accordingly. Since f cyclically permutes the connected components of $\mathscr{A}^{(\nu)}$, all the annuli contained in $\mathscr{A}^{(\nu)}$ are simultaneously non-amphidrome or amphidrome for each $\nu = 1, 2, \ldots, s$. Apply Theorem 2.3 (i) or Theorem 2.4 (i) as the case may be, then

$$f \mid \mathscr{A}^{(\nu)} : \mathscr{A}^{(\nu)} \to \mathscr{A}^{(\nu)}$$

is rel.∂ isotopic to a homeomorphism

$$f' \mid \mathscr{A}^{(\nu)} : \mathscr{A}^{(\nu)} \to \mathscr{A}^{(\nu)}$$

such that, for each annulus $A_j^{(\nu)}$ in $A^{(\nu)}$,

$$(f' \mid \mathscr{A}^{(\nu)})^{r_\nu} : A_j^{(\nu)} \to A_j^{(\nu)}$$

is a linear twist or a special twist, where r_ν denotes the number of the annuli in $\mathscr{A}^{(\nu)}$.

Applying this isotopy for each $\nu = 1, 2, \ldots, s$, we get a pseudo-periodic map

$$f' : \Sigma_g \to \Sigma_g$$

in standard form, which is isotopic to the original

$$f : \Sigma_g \to \Sigma_g.$$

This completes the proof of (i). □

Proof (of (ii)). Suppose we are given two pseudo-periodic maps

$$f, f' : \Sigma_g \to \Sigma_g$$

in standard form. Suppose they are homotopic. We will show the existence of a homeomorphism

$$h : \Sigma_g \to \Sigma_g$$

isotopic to the identity and such that $f = h^{-1} f' h$.

By Theorem 2.2 (i), there exists a homeomorphism

$$g : \Sigma_g \to \Sigma_g$$

which is isotopic to the identity and sends the precise system of cut curves for f to that for f'. Replacing f' by $g^{-1} f' g$, we may assume that the precise system of cut curves is common to f and f'. Let us denote it by $\{C_i\}_{i=1}^r$. Let $\{A_i\}_{i=1}^r$ and $\{A'_i\}_{i=1}^r$ be the systems of annular neighborhoods of $\{C_i\}_{i=1}^r$ which are invariant under the action of f and of f', respectively. Again there exists a homeomorphism

$$g : \Sigma_g \to \Sigma_g$$

which is isotopic to the identity and sends A_i to A'_i. Replacing f' by $g^{-1} f' g$, we may assume that $\{A_i\}_{i=1}^r$ is common to f and f'. Let us denote $\bigcup_{i=1}^r A_i$ by $\mathscr{A}$ and $\Sigma_g - Int(\mathscr{A})$ by $\mathscr{B}$ as before. □

Claim (C). $f \mid \mathscr{B}$ and $f' \mid \mathscr{B}$ are homotopic as maps of pairs:

$$(\mathscr{B}, \partial \mathscr{B}) \to (\mathscr{B}, \partial \mathscr{B}).$$

Proof. We may assume that $\{C_i\}_{i=1}^r$ are closed geodesic with respect to a certain metric Σ_g. Since

$$f, f' : \Sigma_g \to \Sigma_g$$

are homotopic, they are isotopic, [10, 21]. Let

$$f_\tau : \Sigma_g \to \Sigma_g, \quad 0 \le \tau \le 1,$$

be the isotopy with $f_0 = f$ and $f_1 = f'$. This isotopy gives a homeomorphism

$$F : \Sigma_g \times [0,1] \to \Sigma_g \times [0,1]$$

defined by

$$F(p, \tau) = (f_\tau(p), \tau)$$

for

$$(p, \tau) \in \Sigma_p \times [0,1].$$

Since $\Sigma_g \times [0,1]$ is an irreducible 3-manifold (cf. [68]), we can apply an innermost disk argument to achieve

$$F(\mathscr{C} \times [0,1]) = \mathscr{C} \times [0,1].$$

Also by moving F by a rel.∂ isotopy, we may assume

$$F(\mathscr{A} \times [0,1]) = \mathscr{A} \times [0,1].$$

Then
$$F \mid \mathscr{B} \times [0,1] : \mathscr{B} \times [0,1] \to \mathscr{B} \times [0,1]$$
gives a homotopy between $f \mid \mathscr{B}$ and $f' \mid \mathscr{B}$. □

Now we can apply Theorem 2.2, and obtain a homeomorphism
$$h \mid \mathscr{B} : \mathscr{B} \to \mathscr{B}$$
isotopic to the identity $id_{\mathscr{B}}$ and such that
$$f \mid \mathscr{B} = (h \mid \mathscr{B})^{-1}(f' \mid \mathscr{B})(h\mathscr{B}).$$
$h \mid \mathscr{B}$ extends to a homeomorphism
$$h : \Sigma_g \to \Sigma_g$$
which is isotopic to the identity. Replacing f' by $h^{-1} f'h$, we may assume
$$f \mid \mathscr{B} = f' \mid \mathscr{B}.$$

Claim (D). $f \mid \mathscr{A}$ and
$$f' \mid \mathscr{A} : (\mathscr{A}, \partial\mathscr{A}) \to (\mathscr{A}, \partial\mathscr{A})$$
are isotopic by a rel.∂ isotopy.

Proof. Let A_i be an annulus of $\mathscr{A}$. f and f' cause the same permutation on the set of annuli $\{A_i\}_{i=1}^r$ because
$$f \mid \mathscr{B} = f' \mid \mathscr{B}.$$
We have
$$f(A_1) = f'(A_1),$$
which we denote A_2, for simplicity. Let L be a "straight" line in A_1 connecting a point $p_0 \in \partial_0 A_1$ and another $p_1 \in \partial_1 A_1$. Then the images $f(L)$ and $f'(L)$ are arcs in A_2 connecting
$$q_0 := f(p_0) = f'(p_0)$$
and
$$q_1 := f(p_1) = f'(p_1).$$
The arcs $f(L)$ and $f'(L)$ are homotopic in A_2 fixing the end points $\{q_0, q_1\}$. (*Proof.* Note that
$$(f')^{-1} f : \Sigma_g \to \Sigma_g$$
is a pseudo-periodic map, because
$$(f')^{-1} f \mid \mathscr{B} = id_{\mathscr{B}}.$$

If the assertion above is not correct then $(f')^{-1}f$ would have non-zero screw number about C_1, the center line of A_1. But this is impossible by Theorem 2.2, because

$$(f')^{-1}f : \Sigma_g \to \Sigma_g$$

is homotopic to id_{Σ_g} by the assumption of Theorem 2.1(ii)). Then by the ordinary innermost arc argument, together with the Alexander trick,

$$f \mid A_1 : A_1 \to A_2$$

is rel.∂ isotopic to

$$f' \mid A_1 : A_1 \to A_2.$$

Doing the same argument for each A_i, $i = 1, 2, \ldots, r$, we get Claim (B). As we did in the proof of Theorem 2.1 (i), decompose $\mathscr{A}$ into cyclic orbits

$$\mathscr{A}^{(1)} \cup \mathscr{A}^{(2)} \cup \cdots \cup \mathscr{A}^{(s)}$$

under the action of f. Consider an orbit $\mathscr{A}^{(\nu)}$. By the assumption of Theorem 2.1 (ii), $f^{r_\nu} \mid A_j^{(\nu)}$ and

$$(f')^{r_\nu} \mid A_j^{(\nu)} : A_j^{(\nu)} \to A_j^{(\nu)}$$

are simultaneously linear twists or special twists, where $A_j^{(\nu)}$ is an annulus of $\mathscr{A}^{(\nu)}$ and

$$r_\nu = \#(\mathscr{A}^{(\nu)}).$$

Also by Claim (B), $f \mid \mathscr{A}^{(\nu)}$ and

$$f' \mid \mathscr{A}^{(\nu)} : (\mathscr{A}^{(\nu)}, \partial\mathscr{A}^{(\nu)}) \to (\mathscr{A}^{(\nu)}, \partial\mathscr{A}^{(\nu)})$$

are isotopic by a rel.∂ isotopy. Then we can apply Theorem 2.3 (ii) or Theorem 2.4 (ii) as the case may be, and obtain an isotopy

$$h_\tau^{(\nu)} : \mathscr{A}^{(\nu)} \to \mathscr{A}^{(\nu)}, \quad 0 \le \tau \le 1,$$

such that

1. $h_0^{(\nu)} = id$.
2. $(h_\tau^{(\nu)})^{-1}(f' \mid \partial A^{(\nu)})h_\tau^{(\nu)} = f' \mid \partial\mathscr{A}^{(\nu)}(= f \mid \partial\mathscr{A}^{(\nu)})$ on $\partial\mathscr{A}^{(\nu)}$, and
3. $f \mid \mathscr{A}^{(\nu)} = (h_1^{(\nu)})^{-1}(f' \mid \mathscr{A}^{(\nu)})h_1^{(\nu)}$.

The condition (2) above says that

$$h_\tau^{(\nu)} : \mathscr{A}^{(\nu)} \to \mathscr{A}^{(\nu)}$$

is equivariant on $\partial\mathscr{A}^{(\nu)}$ with respect to the action of

$$f' \mid \partial\mathscr{A}^{(\nu)} \ (= f \mid \partial\mathscr{A}^{(\nu)}).$$

Therefore, we can extend the isotopy

$$h_\tau^{(\nu)}, \quad 0 \leq \tau \leq 1,$$

equivariantly into collar neighborhoods $(\partial\mathscr{B}) \times [0, \varepsilon)$ of $\mathscr{B}$ so that, beyond the collar, the extension gives the identity on

$$\mathscr{B} - (\partial\mathscr{B}) \times [0, \varepsilon).$$

Doing the same construction and the extension for each $\nu = 1, 2, \ldots, s$, we get an isotopy

$$h_\tau : \Sigma_g \to \Sigma_g, \quad 0 \leq \tau \leq 1,$$

such that
(a) $h_\tau(\mathscr{A}) = \mathscr{A}$ and $h_\tau(\mathscr{B}) = \mathscr{B}$,
(b) $h_0 = id$,
(c) $h_\tau \mid \mathscr{B} \to \mathscr{B}$ is equivariant with respect to $f' \mid \mathscr{B} = f \mid \mathscr{B}$, i.e.

$$(h_\tau \mid \mathscr{B})^{-1}(f' \mid \mathscr{B})(h_\tau \mid \mathscr{B}) = f' \mid \mathscr{B}(= f \mid \mathscr{B}), \text{ and}$$

(d) $h_\tau \mid \mathscr{A}^{(\nu)} = h_\tau^{(\nu)}, \nu = 1, 2, \ldots, s.$

Now, since

$$f \mid \mathscr{B} = (h_1 \mid \mathscr{B})^{-1}(f' \mid \mathscr{B})(h_1 \mid \mathscr{B})$$

and

$$f \mid \mathscr{A}^{(\nu)} = (h_1 \mid \mathscr{A}^{(\nu)})^{-1}(f' \mid \mathscr{A}^{(\nu)})(h_1 \mid \mathscr{A}^{(\nu)}), \nu = 1, 2, \ldots, s,$$

it is evident that

$$f = h_1^{-1} f' h_1.$$

This completes the proof of Theorem 2.1 (ii). □

Chapter 3
Generalized Quotient

3.1 Definitions and Main Theorem of Chap. 3

A periodic map f on a surface Σ defines a quotient space Σ/f. In the case of a pseudo-periodic map

$$f : \Sigma \to \Sigma,$$

however, the quotient space Σ/f would not be any reasonable space, if the term "quotient space" is taken in the usual sense, i.e., the orbit space under the action of f. To adjust this, we introduce the following definition.

Definition 3.1. A pseudo-periodic map

$$f : \Sigma_g \to \Sigma_g$$

is said to be *of negative twist* if each cut curve C_i has negative screw number in the precise system of cut curves subordinate to f.

Thus, the purpose of this and the next chapters is to generalize the notion of quotient spaces so that a pseudo-periodic map of negative twist always has its "generalized quotient space".

Let us begin by recalling Bers' definition of Riemann surfaces with nodes [11]: A *Riemann surface with nodes*, S, is a connected complex space such that every $p \in S$ has arbitrarily small neighborhoods isomorphic either to the set $|z| < 1$ in $\mathbf{C}$ or to the set $|z| < 1$, $|w| < 1$, $zw = 0$ in $\mathbf{C}^2$. In the second case, p is called a *node*.

When we disregard the complex structure, we will call S a *chorizo space*[1]. A *chorizo space with boundary* is defined in the obvious way. Following Bers [11], we call every component of S-{nodes} a *part* of S. The closure of a part is an *irreducible component*. We call a neighborhood N_p of a node p (isomorphic to

[1] chorizo = Spanish sausage, good to be eaten fried with some wine.

Y. Matsumoto and J.M. Montesinos-Amilibia, *Pseudo-periodic Maps and Degeneration of Riemann Surfaces*, Lecture Notes in Mathematics 2030,
DOI 10.1007/978-3-642-22534-5_3,

the set $|z| < 1, |w| < 1, \ zw = 0$ in $\mathbf{C}^2$) a *nodal neighborhood*. We sometimes consider a *closed nodal neighborhood* $\overline{N}_p$ corresponding to $|z| \leq 1, |w| \leq 1$, $wz = 0$. A nodal neighborhood N_p has two *banks* (cf. [11]) corresponding to $\{|z| < 1, \ w = 0\}$ and $\{z = 0, \ |w| < 1\}$. A connected component of

$$S - \bigcup_{p = \text{node}} N_p$$

is called a *closed part*. (Of course, a closed part may have a non-empty boundary!).

The following definition generalizes (at least in the case when S_1 is non-singular) the notion of a *deformation* $S_1 \to S$ which was introduced by Bers ([11]).

Definition 3.2. Let Σ be a surface, S a chorizo space. A continuous map

$$\pi : \Sigma \to S$$

is called a *pinched covering* if the following conditions are satisfied:

(i) π is surjective,
(ii) for each nodal neighborhood N_p, each connected component A of $\pi^{-1}(N_p)$ is an (open) annulus,

$$\pi|A : A \to N_p$$

maps the central-line C of A to the node p, and

$$\pi|(A - C) : A - C \to N_p - \{p\}$$

is the disjoint union of two cyclic coverings over the two punctured banks, and
(iii) for each part P',

$$\pi|\pi^{-1}(P') : \pi^{-1}(P') \to P'$$

is an orientation-preserving covering (in the usual sense, though $\pi^{-1}(P')$ may not be connected).

In what follows, surfaces and chorizo spaces will be compact. In that case, given a pinched covering

$$\pi : \Sigma \to S,$$

each covering

$$\pi|\pi^{-1}(P') : \pi^{-1}(P') \to P'$$

in it is finite-sheeted, so we can attach this number of sheets to the part P' as its *multiplicity*. We may consider this multiplicity to be attached to the closed part P, which P' contains, or to the irreducible component Θ which contains P', as well.

Definition 3.3. A chorizo space S is *numerical* if to each irreducible component is attached a positive integer (called the *multiciply*).

In a numerical chorizo space, each bank D of a nodal neighborhood N_p gets its multiplicity from the irreducible component to which D belongs.

Henceforth all chorizo spaces will be numerical.

Now we give a definition of a generalized quotient of a pseudo-periodic map

$$f : \Sigma_g \to \Sigma_g.$$

This definition may be considered to be a preliminary one, because it provides a generalized quotient only for a very restricted type of pseudo-periodic maps with negative twist. The final definition will appear almost at the end of the *next* chapter, after a uniqueness theorem is proved.

Definition 3.4. A pinched covering

$$\pi : \Sigma_g \to S$$

is called a *generalized quotient* of a pseudo-periodic map

$$f : \Sigma_g \to \Sigma_g$$

if the following conditions are satisfied:

(i) A system of closed nodal neighborhoods

$$\{\overline{N}_p\}_{p = \text{node}}$$

is fixed in S. For every node p, $\pi^{-1}(\overline{N}_p)$ is preserved by f as a set, i.e.

$$f(\pi^{-1}(\overline{N}_p)) = \pi^{-1}(\overline{N}_p),$$

and for every connected component P_i of

$$S - \bigcup_{p = \text{node}} N_p, \qquad (N_p = Int(\overline{N}_p)),$$

$\pi^{-1}(P_i)$ is also preserved by f as a set i.e.

$$f(\pi^{-1}(P_i)) = \pi^{-1}(P_i).$$

(ii) For each connected component P_i of $S - \bigcup_{p = \text{node}} N_p$,

$$f|\pi^{-1}(P_i) : \pi^{-1}(P_i) \to \pi^{-1}(P_i)$$

is a periodic map of order m_i, m_i being the multiplicity of P_i, and

$$\pi|\pi^{-1}(P_i) : \pi^{-1}(P_i) \to P_i$$

is the quotient map by $f|\pi^{-1}(P_i)$. (NB. By the definition of a pinched covering, $f|\pi^{-1}(P_i)$ has no multiple points).

(iii) For each node p, $\pi^{-1}(\overline{N}_p)$ consists of $m\,(=\gcd(m_1, m_2))$ annuli, where m_1, m_2 are the multiplicities of the banks of $\overline{N}_p$, and

$$f|\pi^{-1}(\overline{N}_p) : \pi^{-1}(\overline{N}_p) \to \pi^{-1}(\overline{N}_p)$$

permutes these annuli cyclically.

(iv) For each annulus A in $\pi^{-1}(\overline{N}_p)$,

$$f^m|A : A \to A$$

is a linear twist (Chap. 2), and its screw number is equal to $-1/n_1 n_2$, where $n_1 = m_1/m, n_2 = m_2/m$.

(v) For each node p, there exist parametrizations of the banks

$$\{z|\,|z| \leq 1\} \to D_1,$$
$$\{z|\,|z| \leq 1\} \to D_2$$

sending O to p, such that if we identify D_i with $\{z|\,|z| \leq 1\}$ through these parametrizations ($i = 1, 2$) and define

$$T_i : D_i \to D_i \quad (i = 1, 2)$$

by

$$T_1(z) = z\exp\left(\frac{\sqrt{-1}\pi}{m_2}(1 - |z|)\right),$$

$$T_2(z) = z\exp\left(\frac{\sqrt{-1}\pi}{m_1}(1 - |z|)\right),$$

then the following identities hold:

$$T_i\pi = \pi f, \quad \text{on } \pi^{-1}(D_i) \quad (i = 1, 2).$$

The definition of a generalized quotient is motivated by the results of Chap. 7.

Remark 3.1. 1. When we talk about a generalized quotient, a system of closed nodal neighborhoods $\{\overline{N}_p\}_{p\,=\,\text{node}}$ will always be fixed in S, so that it satisfies the above conditions, and a closed part P will always be a connected component of

$$S - \bigcup_{p\,=\,\text{node}} N_p,$$

N_p being the interior of $\overline{N}_p$.

2. If

$$\pi : \Sigma_g \to S$$

is a generalized quotient of a pseudo-periodic map

$$f : \Sigma_g \to \Sigma_g,$$

then f is almost in standard form. The difference is that the system of cut curves $\{\pi^{-1}(p)\}_{p=\text{node}}$ may not be a precise system (Chap. 1), but in compensation, the screw numbers are restricted to be as in (iv) above, i.e. they depend on the multiplicities of the banks of p. We will say that such an f is in *superstandard form*.

3. In a generalized quotient

$$\pi : \Sigma_g \to S,$$

every annulus in $\pi^{-1}(\overline{N}_p)$ is non-amphidrome. Amphidrome annuli are hidden in the global configuration of S, in a sense that will be clarified later (see Chap. 4).

Lemma 3.1. *Let*

$$\pi : \Sigma_g \to S$$

be a generalized quotient of a pseudo-periodic map

$$f : \Sigma_g \to \Sigma_g$$

of negative twist. Let $p \in S$ be a node. Let $\overline{N}_p$ be the fixed closed nodal neighborhood of p. Let D_1, D_2 be the two banks of $\overline{N}_p$. Thus, $\overline{N}_p$ is the one point union $D_1 \vee D_2$. Let m_1, m_2 be the multiplicities of D_1, D_2, respectively. Suppose the boundary curves of D_1 and D_2, ∂D_1 and ∂D_2, are oriented as the boundaries of the closed parts P_1, P_2 which are adjacent to $\partial D_1, \partial D_2$, respectively, and let $(\mu_i, \lambda_i, \sigma_i)$ be the valency of the curves in $\pi^{-1}(\partial D_i)$ with respect to the periodic action of

$$f : \pi^{-1}(P_i) \to \pi^{-1}(P_i), \quad i = 1, 2.$$

Then

(i) $\mu_1 = \mu_2 = \gcd(m_1, m_2)$,
(ii) $\lambda_1 = m_1/\gcd(m_1, m_2)$, $\lambda_2 = m_2/\gcd(m_1, m_2)$, *(i.e. $\lambda_1 = n_1, \lambda_2 = n_2$, in the notation of the definition of a generalized quotient, (iv)), and*
(iii) $\sigma_1 \equiv n_2 \pmod{n_1}$, $\sigma_2 \equiv n_1 \pmod{n_2}$.

Proof. (i) Since

$$f : \pi^{-1}(N_p) \to \pi^{-1}(N_p)$$

causes a cyclic permutation of the components, the numbers μ_1, μ_2 are equal to the number of the annuli contained in $\pi^{-1}(N_p)$, which equals $\gcd(m_1, m_2)$ by the definition of a generalized quotient. (See the proof of Corollary 2.1, (i)).

(ii) Remember that the multiplicity m_i is the order of the periodic map

$$f|\pi^{-1}(P_i) : \pi^{-1}(P_i) \to \pi^{-1}(P_i).$$

Then $f|\pi^{-1}(\partial D_i)$ has order m_i, while by (i) $\pi^{-1}(\partial D_i)$ contains m components, m being $\gcd(m_1, m_2)$. Thus for each component $\partial\tilde{D}_i$ of $\pi^{-1}(\partial D_i)$,

$$f^m : \partial\tilde{D}_i \to \partial\tilde{D}_i$$

has order m_i/m. This is λ_i by definition.

(iii) Let δ_1 and δ_2 be defined by

$$\sigma_1\delta_1 \equiv 1 \pmod{\lambda_1}, \quad 0 \le \delta_1 < \lambda_1,$$
$$\sigma_2\delta_2 \equiv 1 \pmod{\lambda_2}, \quad 0 \le \delta_2 < \lambda_2.$$

Applying Corollary 2.1 (ii) to the disjoint union of annuli $\pi^{-1}(\overline{N}_p)$ whose members have screw number $-1/n_1n_2$, we have

$$-\frac{1}{n_1n_2} + \frac{\delta_1}{n_1} + \frac{\delta_2}{n_2} = \alpha \quad \text{an integer}$$

(Remember $\lambda_1 = n_1$, $\lambda_2 = n_2$.) Then

$$n_2\delta_1 + n_1\delta_2 = 1 + \alpha n_1 n_2,$$

from which we have

$$n_2\delta_1 \equiv 1 \pmod{n_1}, \quad \text{and} \quad n_1\delta_2 \equiv 1 \pmod{n_2}.$$

Since $\gcd(n_1, n_2) = 1$, we obtain

$$\sigma_1 \equiv n_2 \pmod{n_1}, \quad \text{and} \quad \sigma_2 \equiv n_1 \pmod{n_2}.$$ □

Proposition 3.1. *Let*

$$\pi : \Sigma_g \to S$$

be a generalized quotient of a pseudo-periodic map

$$f : \Sigma_g \to \Sigma_g.$$

Let Θ_0 be an irreducible component of S with multiplicity m_0. Let

$$\{p_1, p_2, \ldots, p_k\}$$

be the set of the intersection points of Θ_0 with the other irreducible components. Let m_i be the multiplicity of the irreducible component which intersects Θ_0 at

$$p_i (i = 1, 2, \ldots, k).$$

Then

$$m_1 + m_2 + \cdots + m_k$$

is divisible by m_0.

Proof. We add the self-intersection points of Θ_0,

$$\{p_{k+1}, \ldots, p_l\},$$

to

$$\{p_1, p_2, \ldots, p_k\},$$

and set $m_i = m_0$ for $i = k+1, \ldots, l$. Let P be the closed part contained in Θ_0:

$$P = \Theta_0 - \bigcup_{i=1}^{l} Int(D_i),$$

where $D_i = \Theta_0 \cap \overline{N}_{p_i}$. We denote ∂D_i by $\partial_i P$. Let

$$\omega : H_1(P) \to \mathbf{Z}/m_0$$

be the monodromy exponent associated with

$$\pi^{-1}(P) \to \pi^{-1}(P)/f = P.$$

(See the outlined proof of Theorem 1.3 for the definition. Also see [51]). Let $(\mu_i, \lambda_i, \sigma_i)$ be the valency of $\partial_i P$ (with respect to the periodic map

$$f|\pi^{-1}(P) : \pi^{-1}(P) \to \pi^{-1}(P);$$

all curves in the preimage of $\partial_i P$ have the same valency, and, as we remarked in Chap. 1, this is, by definition, the valency of $\partial_i P$ with respect to $f|\pi^{-1}(P)$).

Then, by definition of valency [51, Sect. 2],

$$\omega([\partial_i P]) \equiv \mu_i \, \sigma_i \pmod{m_0},$$

By Lemma 3.1 (i), (iii), we have $\mu_i = \gcd(m_0, m_i)$ and $\sigma_i \equiv m_i/\gcd(m_0, m_i)$ $(\text{mod } (m_0)/\gcd(m_0, m_i))$.

Thus

$$\mu_i \, \sigma_i \equiv m_i \pmod{m_0}.$$

Since

$$\sum_{i=1}^{l} [\partial P_i] = [\partial P] = 0$$

in $H_1(P)$, we have

$$\sum_{i=1}^{l} m_i = \sum_{i=1}^{l} \omega([\partial_i P]) \equiv 0 \pmod{m_0}.$$

Remember that $m_i = m_0$ for $i = k+1, \ldots, l$. Then

$$\sum_{i=1}^{k} m_i \equiv 0 \pmod{m_0}.$$

(Cf. [51, (4.6)]). □

Definition 3.5. A generalized quotient $\pi : \Sigma \to S$ is called a *minimal quotient* if the multiplicities of S satisfy the following *minimality conditions*:

(i) if an irreducible component Θ_0 is a sphere (without self-intersection) and intersects only one other component Θ_i, in one point exactly, then $m_i/m_0 \geq 2$ (NB. m_i/m_0 is an integer by Proposition 3.1), and
(ii) if an irreducible component Θ_0 is a sphere (without self-intersection) and intersects the union of the other components exactly in two points $\{p_1, p_2\}$, then $(m_1 + m_2)/m_0 \geq 2$. (NB. $(m_1 + m_2)/m_0$ is also an integer).

Our main result in Chap. 3 is the following:

Theorem 3.1 (Existence). *Let*

$$f : \Sigma_g \to \Sigma_g$$

be a pseudo-periodic map of negative twist. Then f is isotopic to a pseudo-periodic map

$$f' : \Sigma_g \to \Sigma_g$$

whose minimal quotient

$$\pi : \Sigma_g \to S$$

exists.

Uniqueness of the minimal quotient will be proved in Chap. 4.

3.2 Proof of Theorem 3.1

By Theorem 2.1, a pseudo-periodic map

$$f : \Sigma_g \to \Sigma_g$$

is isotopic to a pseudo-periodic map in standard form. Thus we may assume that f is already in standard form.

Let $\{A_i\}_{i=1}^r$ be the annular neighborhood system of precise cut curves $\{C_i\}_{i=1}^r$ (required in the definition of standard form in Chap. 2) such that

(a) $f(\mathscr{A}) = \mathscr{A}$ and $f(\mathscr{B}) = \mathscr{B}$, where $\mathscr{A} = \bigcup_{i=1}^r A_i$ and $\mathscr{B} = \Sigma_g - Int(\mathscr{A})$,
(b) $f|\mathscr{B} : \mathscr{B} \to \mathscr{B}$ is a periodic map, and
(c) f causes linear twists or special twists, or both, in $\mathscr{A}$.

Let $\mathscr{D}$ be a disjoint union of invariant disk neighborhoods of all the multiple points in $\mathscr{B}$, and let $\mathscr{B}'$ denote $\mathscr{B} - Int(\mathscr{D})$.

Minimal quotients will be constructed for $(\mathscr{A}, f|\mathscr{A})$, $(\mathscr{B}', f|\mathscr{B}')$, and $(\mathscr{D}, f|\mathscr{D})$, separately.

3.3 Quotient of $(\mathscr{B}', f|\mathscr{B}')$

This is the easiest. We have only to take the quotient

$$\pi|\mathscr{B}' : \mathscr{B}' \to \mathscr{B}'/(f|\mathscr{B}')$$

by the periodic action

$$f|\mathscr{B}' : \mathscr{B}' \to \mathscr{B}'.$$

Since we have deleted all the multiple points, $\mathscr{B}'/(f|\mathscr{B}')$ is a surface free from cone points. Each connected component P of $\mathscr{B}'/(f|\mathscr{B}')$ acquires its multiplicity as the number of the sheets over it. By formal reasons, we denote the numerical surface $\mathscr{B}'/(f|\mathscr{B}')$ by $Ch(\mathscr{B}')$ and the quotient by

$$\pi : \mathscr{B}' \to Ch(\mathscr{B}').$$

3.4 Re-normalization of a Rotation

We will construct a generalized quotient for $(\mathscr{D}, f|\mathscr{D})$. First, let us consider the case in which f permutes the disks of $\mathscr{D}$ cyclically.

Assume

$$\mathscr{D} = \bigcup_{\alpha=1}^{m} \Delta_\alpha$$

and

$$f(\Delta_\alpha) = \Delta_{\alpha+1}, \quad \text{for } \alpha = 1, 2, \ldots, m-1, \quad \text{and} \quad f(\Delta_m) = \Delta_1.$$

Also assume that

$$f^m : \Delta_1 \to \Delta_1$$

is a rotation of angle $2\pi\, \delta/\lambda$, where δ, λ are integers satisfying

$$\gcd(\delta,\, \lambda) = 1, \quad 0 < \delta < \lambda.$$

Then the same rotation appears in the other disks

$$\Delta_\alpha, \quad \alpha = 2,\, \ldots,\, m.$$

Let σ be defined by

$$\sigma\delta \equiv 1 \pmod{\lambda}, \quad \text{and} \quad 0 < \sigma < \lambda.$$

Note that (m, λ, σ) is the valency of the centers of the Δ_α's.
Now let

$$(n_0,\, n_1,\, \ldots,\, n_l), \quad l \geq 1,$$

be a sequence of positive integers satisfying

(α) $n_0 = \lambda,\ n_1 = \lambda - \sigma$,
(β) $n_0 > n_1 > \cdots > n_l = 1$, and
(γ) $n_{i-1} + n_{i+1} \equiv 0 \pmod{n_i}$, $i = 1, 2, \ldots, l-1$.

Remark 3.2.

1. Condition (γ) is empty, if $l = 1$, i.e., if $\lambda - \sigma = 1$.
2. Conditions (β), (γ) imply

$$\gcd(n_0,\, n_1) = \gcd(n_1,\, n_2) = \cdots = \gcd(n_{l-1},\, n_l) = 1$$

3. Since $n_{i-1} > n_i$, we have

$$(n_{i-1} + n_{i+1})/n_i \geq 2,$$

for $i = 1, 2, \ldots, l-1$.
4. Integers

$$n_0,\, n_1,\, \ldots,\, n_l$$

satisfying (α), (β), (γ) exist uniquely by the Euclidean algorithm.

Claim (E). Let $(n_0,\, n_1,\, \ldots,\, n_l)$ be the above integers. Then for each k with $0 < k < l$ we have

$$\sum_{i=k}^{l-1} \frac{1}{n_i n_{i+1}} = \frac{n_k - \delta_k}{n_k} \quad ; \text{for some integer } \delta_k,$$

where δ_k satisfies

$$\gcd(\delta_k,\, n_k) = 1, \quad 0 < \delta_k < n_k,$$

and for $k = 0$ we have

$$\sum_{i=0}^{l-1} \frac{1}{n_i\, n_{i+1}} = \frac{\lambda - \delta}{\lambda}.$$

This claim is nothing but Lemma 5.2. (3), which will be proved in Chap. 5.

Remark 3.3. Ashikaga and Ishizaka [9] noticed that the formula of Claim (E) can be used to prove the reciprocity law of the Dedekind sum. See also [25, 47].

Consider Δ_1 to be a unit disk in **C** and divide it by $2l$ concentric circles

$$\Gamma_{2l-1},\ \Gamma_{2l-2},\ \ldots,\ \Gamma_1,\ \Gamma_0$$

of radii

$$1/2l,\ 2/2l,\ \ldots,\ (2l-1)/2l,\ 1,$$

into $2l - 1$ annuli and a disk (NB. $\Gamma_0 = \partial\Delta_1$). Let Δ_1' denote the central disk bounded by Γ_{2l-1}, and let A_i denote the annulus between Γ_i and Γ_{i-1}.

For convenience, A_i is called a "black" annulus if i is odd and a "white" annulus if i is even.

We identify the disk Δ_α with Δ_1 through

$$f^{\alpha-1} : \Delta_1 \to \Delta_\alpha, \quad \alpha = 2, \ldots, m,$$

and we will deform the last step

$$f : \Delta_m \to \Delta_1$$

by a *rel.∂ isotopy* so that, if

$$f' : \Delta_m \to \Delta_1$$

denotes the resulting homeomorphism, then the composition

$$f' f^{m-1} : \Delta_1 \to \Delta_1$$

satisfies the following conditions:

(i) $f' f^{m-1}$ preserves

$$\Gamma_{2l-1},\ \Gamma_{2l-2},\ \ldots,\ \Gamma_1,\ \Gamma_0,$$

(ii) on a white annulus A_{2k},

$$f' f^{m-1} : A_{2k} \to A_{2k}$$

is a rotation of angle

$$2\pi\, \delta_k / n_k, \quad k = 1, 2, \dots, l-1,$$

δ_k, n_k being as in Claim (E),

(iii) on the central disk Δ'_1,

$$f' f^{m-1} : \Delta'_1 \to \Delta'_1$$

is the identity, and

(iv) on a black annulus A_{2k-1},

$$f' f^{m-1} : A_{2k-1} \to A_{2k-1}$$

is a linear twist with screw number

$$-1/n_{k-1} n_k, \quad k = 1, 2, \dots, l.$$

(Twists occur only in black annuli.)

Example 3.1 (Fig. 3.1). Set $\lambda = 10$, $\delta = 3$. Then $\sigma = 7$.
$n_0 = 10$, $n_1 = 10 - 7 = 3$, $n_2 = 2$, $n_3 = 1$, where $l = 3$.

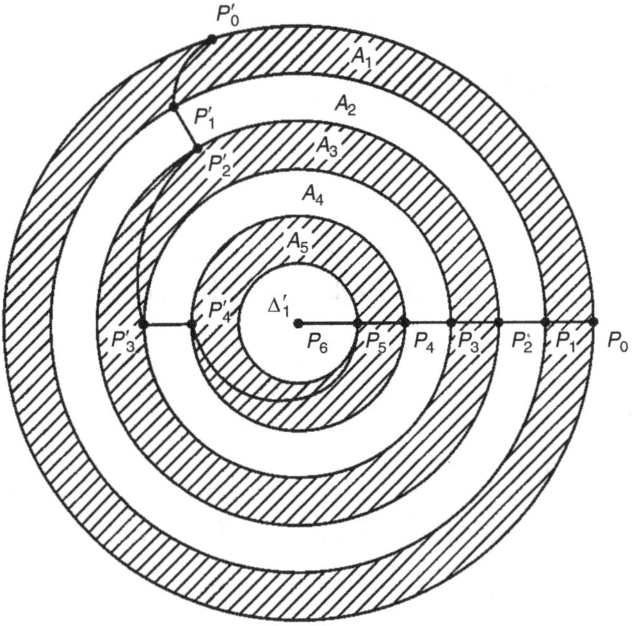

Fig. 3.1 $f' f^{m-1} : \Delta_1 \to \Delta_1$ maps the arc $P_0 P_1 P_2 P_3 P_4 P_5 P_6$ onto the arc $P'_0 P'_1 P'_2 P'_3 P'_4 P_5 P_6$

$$\frac{1}{n_2 n_3} = \frac{1}{2}, \quad \delta_2 = 1.$$

$$\frac{1}{n_1 n_2} + \frac{1}{n_2 n_3} = \frac{1}{3 \cdot 2} + \frac{1}{2} = \frac{2}{3}, \quad \delta_1 = 1$$

$$\frac{1}{n_0 n_1} + \frac{1}{n_1 n_2} + \frac{1}{n_2 n_3} = \frac{1}{10 \cdot 3} + \frac{1}{3 \cdot 2} + \frac{1}{2} = \frac{7}{10}, \quad \delta_0 = \delta = 3.$$

In the general case, the picture is essentially the same as in this example. We first define $f' f^{m-1}$ on the white annuli (or on Δ_1') to be the rotations prescribed by Condition (ii) (or as the identity). Then the screw numbers in the black annuli are *automatically* adjusted as prescribed by Condition (iv), because of Claim (E). (Of course, we must avoid "artificial" full Dehn twists.) On each black annulus, this twist is *rel.*∂ isotopic to a linear twist by Theorem 2.3. (See also Lemma 2.1.) Thus we get the required

$$f' : \Delta_m \to \Delta_1.$$

Define

$$f' : \mathscr{D} \to \mathscr{D}$$

by setting

$$f'|\Delta_\alpha = f|\Delta_\alpha, \quad \alpha = 1, 2, \ldots, m-1, \qquad \text{and}$$
$$f'|\Delta_m = \text{the above } f' : \Delta_m \to \Delta_1.$$

Then

$$f : \mathscr{D} \to \mathscr{D}$$

is *rel.*∂ isotopic to

$$f' : \mathscr{D} \to \mathscr{D},$$

and in each Δ_α,

$$(f')^m : \Delta_\alpha \to \Delta_\alpha$$

behaves just as described by conditions (i), (ii), (iii) and (iv).

This completes the *re-normalization* of

$$f : \mathscr{D} \to \mathscr{D},$$

and we say that the re-normalization

$$f' : \mathscr{D} \to \mathscr{D}$$

is in *superstandard form.*

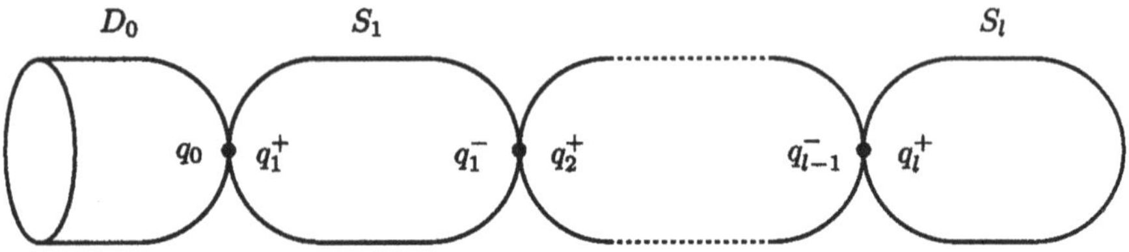

Fig. 3.2 Chorizo space $Ch(\mathscr{D})$

Next we will construct a chorizo space $Ch(\mathscr{D})$ and a pinched covering

$$\pi : \mathscr{D} \to Ch(\mathscr{D})$$

so that $\pi : \mathscr{D} \to Ch(\mathscr{D})$ is a generalized quotient of the re-normalized $f' : \mathscr{D} \to \mathscr{D}$.

Take a disk D_0 and l spheres

$$S_1, S_2, \ldots, S_l.$$

Let q_0 be the center of D_0, and q_i^+ (resp. q_i^-) the north (resp. the south) pole of S_i, $i = 1, 2, \ldots, l$. Identify q_0 with q_1^+, and q_i^- with q_{i+1}^+ for $i = 1, 2, \ldots, l-1$. Then we have our (not yet numerical) chorizo space $Ch(\mathscr{D})$. See Fig. 3.2.

The pinched covering

$$\pi : \mathscr{D} \to Ch(\mathscr{D})$$

is constructed as follows. First we will define

$$\pi : \Delta_1 \to Ch(\mathscr{D}).$$

Let $C_k (\subset \Delta_1)$ be the center-line of the black annulus A_{2k-1} $(k = 1, 2, \ldots, l)$. The complement, $A_{2k-1} - C_k$ consists of two half-open annuli, A'_{2k-1} and A''_{2k-1}, where A'_{2k-1} denotes the outer one. Then the pinched covering

$$\pi : \Delta_1 \to Ch(\mathscr{D})$$

is defined by setting

1. $\pi(C_k) = q_k^+, \quad k = 1,\ 2,\ \ldots,\ l,$
2. the map

$$\pi|(A''_{2k-1} \cup A_{2k} \cup A'_{2k+1}) : A''_{2k-1} \cup A_{2k} \cup A'_{2k+1} \to S_k - \{q_k^+, q_k^-\}$$

is an n_k-fold cyclic covering whose covering translations are generated by the rotation of angle $2\pi/n_k$.

3. the map

$$\pi|A_1' : A_1' \to D_0 - \{q_0\}$$

is an $n_0(= \lambda)$-fold cyclic covering whose covering translations are generated by the rotation of angle $2\pi/n_0$.
4. the map

$$\pi|\Delta_1' \cup A_{2l-1}'' : \Delta_1' \cup A_{2l-1}'' \to S_l - \{q_l^+\}$$

is a homeomorphism.

These conditions determine a pinched covering

$$\pi : \Delta_1 \to Ch(\mathscr{D}).$$

The image $\pi(A_{2k-1})$ of a black annulus A_{2k-1} is a closed nodal neighborhood $\overline{N}_k$ of q_k^+ consisting of two banks

$$\pi(A_{2k-1}' \cup C_k) \ \ (=: D_{k-1}^-)$$

and

$$\pi(A_{2k-1}'' \cup C_k) \ \ (=: D_k^+).$$

Before constructing the projection

$$\pi : \Delta_\alpha \to Ch(\mathscr{D})$$

for the other disks

$$\Delta_\alpha, \quad \alpha = 2, \ldots, m,$$

we must parameterize the two banks, D_k^+, D_{k-1}^-, for each $k = 1, \ldots, l$. The parametrizations

$$\{z|\,|z| \leq 1\} \to D_k^+, \ \ D_{k-1}^-$$

are chosen so that, if we identify D_k^+ and D_{k-1}^- with $\{z|\,|z| \leq 1\}$ through these parametrizations,

$$\pi|\Delta_1 : A_{2k-1}' \cup C_k \to D_{k-1}^-$$

is described as

$$\pi\left(r\exp(\sqrt{-1}\theta)\right) = \frac{r-b}{a-b}\exp(\sqrt{-1}n_{k-1}\theta)$$

and

$$\pi|\Delta_1 : A''_{2k-1} \cup C_k \to D_k^+$$

is described as

$$\pi\left(r\exp(\sqrt{-1}\theta)\right) = \frac{b-r}{b-c}\exp(-\sqrt{-1}n_k\theta),$$

where the annuli $A'_{2k-1} \cup C_k$ and $A''_{2k-1} \cup C_k$ are regarded, for simplicity, as

$$A'_{2k-1} \cup C_k = \left\{ r\exp(\sqrt{-1}\theta) \mid b \le r \le a,\ \ 0 \le \theta \le 2\pi \right\},$$

and

$$A''_{2k-1} \cup C_k = \left\{ r\exp(\sqrt{-1}\theta) \mid c \le r \le b,\ \ 0 \le \theta \le 2\pi \right\}.$$

It is easy to verify that with these parametrizations of D_k^+, D_{k-1}^- the following identities hold:

$$(T_{k-1}^-)^m(\pi|\Delta_1) = (\pi|\Delta_1)(f')^m \quad \text{on} \quad A'_{2k-1} \cup C_k$$

and

$$(T_k^+)^m(\pi|\Delta_1) = (\pi|\Delta_1)(f')^m \quad \text{on} \quad A''_{2k-1} \cup C_k$$

where

$$T_{k-1}^- : D_{k-1}^- \to D_{k-1}^-$$

and

$$T_k^+ : D_k^+ \to D_k^+$$

are defined by

$$T_{k-1}^-(z) = z\exp\left(\frac{\sqrt{-1}\pi}{mn_k}(1-|z|)\right),$$

and

$$T_k^+(z) = z\exp\left(\frac{\sqrt{-1}\pi}{mn_{k-1}}(1-|z|)\right),$$

respectively.

Now we are in a position to construct

$$\pi : \Delta_\alpha \to Ch(\mathscr{D}), \quad \text{for} \ \ \alpha = 2, \ \ldots, \ m.$$

The definition is as follows:

$$\pi|\Delta_\alpha = \begin{cases} (T_{k-1}^-)^{\alpha-1}(\pi|\Delta_1)[(f)^{\alpha-1}]^{-1} & \text{on} \ \ f^{\alpha-1}(A'_{2k-1} \cup C_k) \ (k = 1, 2, ..., l), \\ (T_k^+)^{\alpha-1}(\pi|\Delta_1)[(f)^{\alpha-1}]^{-1} & \text{on} \ \ f^{\alpha-1}(A''_{2k-1} \cup C_k) \ (k = 1, 2, ..., l), \\ (\pi|\Delta_1)[(f)^{\alpha-1}]^{-1} \ \text{on} \ f^{\alpha-1}(A_{2k}) & \text{and} \ \ f^{\alpha-1}(\Delta'_1) \ (k = 1, 2, ..., l-1). \end{cases}$$

Taking the disjoint union, we obtain the pinched covering

$$\pi = \bigcup_{\alpha=1}^{m} (\pi|\Delta_\alpha) : \mathscr{D} \to Ch(\mathscr{D}).$$

Let P_k denote the closed part

$$S_k - (Int(D_k^+) \cup Int(D_k^-)).$$

The re-normalization $f' : \mathscr{D} \to \mathscr{D}$ preserves $\pi^{-1}(\overline{N}_k)$ and $\pi^{-1}(P_k)$ (condition (i) of generalized quotients). For each P_k,

$$f' : \pi^{-1}(P_k) \to \pi^{-1}(P_k)$$

is a periodic map of order mn_k, and

$$\pi : \pi^{-1}(P_k) \to P_k$$

is its projection (condition (ii)). For each node q_k^+, the number of annuli in $\pi^{-1}(\overline{N}_k)$ is m, and

$$f' : \pi^{-1}(\overline{N}_k) \to \pi^{-1}(\overline{N}_k)$$

cyclically permutes the components (condition (iii)).

$$f' : \pi^{-1}(\overline{N}_k) \to \pi^{-1}(\overline{N}_k)$$

is a linear twist in each annulus, and the screw number is $-1/n_{k-1}n_k$ (condition (iv)). Finally condition (v) is easily verified by the construction above. Thus attaching the multiplicities

$$mn_0, mn_1, \ldots, mn_l \ (= m)$$

to the components

$$D_0, S_1, \ldots, S_l,$$

we can make $Ch(\mathscr{D})$ a numerical chorizo, and $\pi : \mathscr{D} \to Ch(\mathscr{D})$ a generalized quotient of f'.

The above argument can be summarized as follows:

Lemma 3.2. *Let $\mathscr{D}$ be a disjoint union of m disks,*

$$\Delta_1,\ \Delta_2,\ \ldots,\ \Delta_m.$$

Suppose that a homeomorphism $f : \mathscr{D} \to \mathscr{D}$ permutes these disks cyclically, and that for each α $(= 1, 2, \ldots, m)$,

$$f^m : \Delta_\alpha \to \Delta_\alpha$$

is a rotation of angle $2\pi\delta/\lambda$, where δ, λ are integers satisfying

$$\gcd(\delta,\ \lambda) = 1, \quad 0 < \delta < \lambda.$$

Let σ be an integer such that

$$\sigma\delta \equiv 1 \pmod{\lambda}$$

and

$$0 < \sigma < \lambda$$

(meaning that the centers of the Δ_α's have the valency (m, λ, σ)). *Let*

$$(n_0,\ n_1,\ \ldots,\ n_l), \quad l \geq 1,$$

be the sequence of positive integers determined by

$$n_0 = \lambda,\ n_1 = \lambda - \sigma,$$

$$n_0 > n_1 > \cdots > n_l = 1$$

and

$$n_{i-1} + n_{i+1} \equiv 0 \pmod{n_i}, \quad i = 1,\ 2,\ \ldots,\ l-1.$$

Let $Ch(\mathscr{D})$ be the chorizo space as shown in Fig. 3.2, *and attach the multiplicities*

$$mn_0,\ mn_1,\ \ldots,\ mn_l\ (= m),$$

to the components

$$D_0,\ S_1,\ \ldots,\ S_l.$$

Then there exist a pinched covering

$$\pi : \mathscr{D} \to Ch(\mathscr{D})$$

and a rel.∂ isotopy from

$$f : \mathscr{D} \to \mathscr{D}$$

to

$$f' : \mathscr{D} \to \mathscr{D}$$

such that

$$\pi : \mathscr{D} \to Ch(\mathscr{D})$$

is a generalized quotient of f'.

Remark 3.4. 1. By Remark 3.2, 3) before Claim (E), the sphere components

$$S_1,\ S_2,\ \ldots,\ S_l$$

of $Ch(D)$ satisfy the minimality condition.
2. The passage from the cone D/f to the chorizo space $Ch(D)$ corresponds to blowing up the cone point (see [33] for a description of the blowing up process).

In the general case when $f : \mathscr{D} \to \mathscr{D}$ does not necessarily permute the disks cyclically, decompose the disks into cyclic orbits :

$$\mathscr{D} = \mathscr{D}^{(1)} \cup \mathscr{D}^{(2)} \cup \ldots \cup \mathscr{D}^{(s)},$$

and define

$$\pi : \mathscr{D} \to Ch(\mathscr{D})$$

to be the disjoint union

$$\bigcup_{\nu=1}^{s} \mathscr{D}^{(\nu)} \to \bigcup_{\nu=1}^{s} Ch(\mathscr{D}^{(\nu)}).$$

3.5 Re-normalization of a Linear Twist

Let

$$\mathscr{A} = \bigcup_{\alpha=1}^{m} A_\alpha$$

be a disjoint union of annuli. Suppose that a homeomorphism

$$f : \mathscr{A} \to \mathscr{A}$$

cyclically permutes the components, and that

$$f^m : A_\alpha \to A_\alpha$$

is a linear twist (of *negative* screw number) for each $\alpha = 1,\ 2,\ \ldots,\ m.$

We will construct a chorizo space $Ch(\mathscr{A})$ and a pinched covering

$$\pi : \mathscr{A} \to CH(\mathscr{A}),$$

and will show that

$$f : \mathscr{A} \to \mathscr{A}$$

is *rel.*∂ isotopic to a homeomorphism

$$f' : \mathscr{A} \to \mathscr{A}$$

such that

$$\pi : \mathscr{A} \to Ch(\mathscr{A})$$

is a generalized quotient of f'.

We may assume

$$f(A_\alpha) = A_{\alpha+1}, \quad \alpha = 1, \ldots, m-1,$$

and $f(A_m) = A_1$. Let us identify A_1 with $[0, 1] \times S^1$ (as we did in Chap. 2), S^1 being parametrized as $\mathbf{R}/\mathbf{Z}$.

Let

$$(m,\ \lambda_0,\ \sigma_0)$$

and

$$(m,\ \lambda_1,\ \sigma_1)$$

be the valencies of the boundary curves

$$\partial_0 A_1 = \{0\} \times S^1$$

and

$$\partial_1 A_1 = \{1\} \times S^1$$

with respect to

$$f : \mathscr{A} \to \mathscr{A}.$$

We introduce integers δ_0, δ_1 as usual:

$$\sigma_0\, \delta_0 \equiv 1 \pmod{\lambda_0}, \quad 0 \leq \delta_0 < \lambda_0,$$
$$\sigma_1\, \delta_1 \equiv 1 \pmod{\lambda_1}, \quad 0 \leq \delta_1 < \lambda_1;$$

($\delta_0 = 0$ iff $\lambda_0 = 1$. Similarly for δ_1). Finally let s be the screw number of

$$f : \mathscr{A} \to \mathscr{A}$$

in

$$A_\alpha, \quad \alpha = 1, \dots, m,$$

which is independent of α. Note that $s < 0$ by our assumption.

By Corollary 2.1,

$$-|s| + \delta_0/\lambda_0 + \delta_1/\lambda_1$$

is an integer.

Claim (F).

1. There exists uniquely a sequence of positive integers

$$(n_0, n_1, \dots, n_l), \quad l \geq 1,$$

satisfying the following conditions $(\alpha) \sim (\varepsilon)$:

(α) $n_0 = \lambda_0, \ n_l = \lambda_1,$
(β) $n_1 \equiv \sigma_0 \pmod{\lambda_0}, \quad n_{l-1} \equiv \sigma_1 \pmod{\lambda_1},$
(γ) $n_{i-1} + n_{i+1} \equiv 0 \pmod{n_i}, \quad i = 1, 2, \dots, l-1,$
(δ) $(n_{i-1} + n_{i+1})/n_i \geq 2, \quad i = 1, 2, \dots, l-1,$ and
(ε) $\sum_{i=0}^{l-1} 1/(n_i n_{i+1}) = |s|.$

2. The above sequence $(n_0, n_1, \dots, n_l)$ has additional properties:

(ζ) for each k with $0 < k < l$, we have

$$-\frac{\delta_0}{\lambda_0} + \sum_{i=0}^{k-1} \frac{1}{n_i\, n_{i+1}} = \frac{d_k}{n_k}$$

for some integer d_k coprime with n_k, and

(η)

$$-\frac{\delta_0}{\lambda_0} + \sum_{i=0}^{l-1} \frac{1}{n_i\, n_{i+1}} \equiv \frac{\delta_1}{\lambda_1} \pmod{1}.$$

Claim (F), (1) is nothing but Theorem 5.1, which will be proved in Chap. 5. Claim (F), (2) (ζ) is Lemma 5.2 (1). The last property, (η), follows from (ε) and the fact that

$$-|s| + \delta_0/\lambda_0 + \delta_1/\lambda_1$$

is an integer.

Note that

$$\gcd(n_i, n_{i+1}) = 1$$

for $i = 0, \dots, l-1$ because of (α), (β) and (γ).

Let us proceed taking the above claim for granted.

Divide $A_1 = [0, 1] \times S^1$ by $2(l - 1)$ circles

$$\Gamma_i = \{i/(2l - 1)\} \times S^1, \quad i = 1, 2, \ldots, 2l - 2$$

into $2l - 1$ annuli. We set

$$\Gamma_0 = \partial_0 A_1 = \{0\} \times S^1$$

and

$$\Gamma_{2l-1} = \partial_1 A_1 = \{1\} \times S^1.$$

Let Z_i denote the annulus between Γ_{i-1} and Γ_i.

As in the case of a rotation, we call Z_i a "black" annulus if i is odd and a "white" annulus if i is even.

We identify the α-th annulus A_α with A_1 through

$$f^{\alpha-1} : A_1 \to A_\alpha, \quad \alpha = 2, \ldots, m,$$

and will deform the last step

$$f : A_m \to A_1$$

by a rel.∂ isotopy so that the resulting

$$f' : A_m \to A_1$$

satisfies the following conditions:

(i) The map

$$f' f^{m-1} : A_1 \to A_1$$

preserves

$$\Gamma_0, \Gamma_1, \ldots, \Gamma_{2l-1},$$

(ii) on a white annulus Z_{2k},

$$f' f^{m-1} : Z_{2k} \to Z_{2k}$$

is a "rotation" sending (t, x) to

$$(t, x + d_k/n_k), \quad k = 1, 2, \ldots, l - 1,$$

d_k, n_k being as in Claim (F), (ζ),

(iii) on a black annulus Z_{2k-1},

$$f' f^{m-1} : Z_{2k-1} \to Z_{2k-1}$$

is a linear twist with screw number

$$-1/(n_{k-1} n_k), \quad k = 1, \dots, l,$$

and

(iv) on $\Gamma_0 = \partial_0 A_1$,

$$f' f^{m-1} = f^m$$

is a rotation sending

$$(0, x) \quad \text{to} \quad (0, x - \delta_0/\lambda_0),$$

and on $\Gamma_1 = \partial_1 A_1$,

$$f' f^{m-1} = f^m$$

is a rotation sending

$$(1, x) \quad \text{to} \quad (1, x + \delta_1/\lambda_1).$$

The last condition (iv) is nothing but the definition of the valencies

$$(m, \lambda_0, \sigma_0)$$

and

$$(m, \lambda_1, \sigma_1).$$

(Remember the orientation conventions in Chap. 2. See also Corollary 2.2.) We add (iv) just as a remark.

Example 3.2 (Fig. 3.3). Set

$$\lambda_0 = 5, \quad \sigma_0 = 2, \quad \lambda_1 = 4, \quad \sigma_1 = 3.$$

Then

$$\delta_0 = 3, \quad \delta_1 = 3.$$

We assume

$$s = -7/20.$$

(NB. $-|s| + \delta_0/\lambda_0 + \delta_1/\lambda_1 = -7/20 + 3/5 + 3/4 = 1$).

Then

$$(n_0, n_1, n_2, n_3) = (5, 2, 3, 4)$$

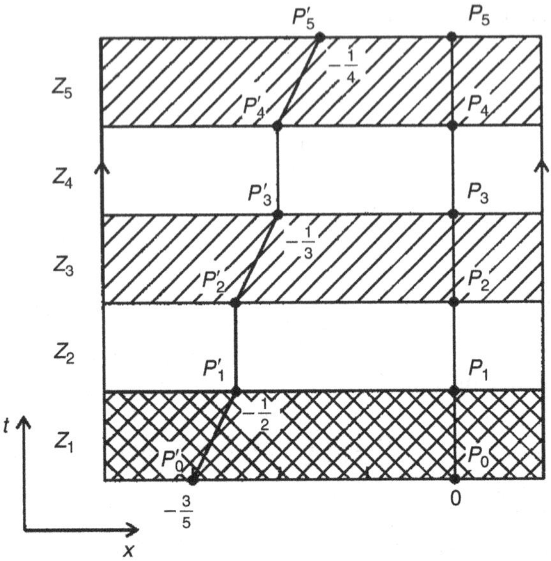

Fig. 3.3 $f' f^{m-1} : A_1 \to A_1$ maps the arc $P_0 P_1 P_2 P_3 P_4 P_5$ onto the arc $P'_0 P'_1 P'_2 P'_3 P'_4 P'_5$

and

$$l = 3.$$

(See the proof of Theorem 5.1 in which there is an algorithm to construct these numbers.)

$$-\frac{\delta_0}{\lambda_0} + \frac{1}{n_0 n_1} = -\frac{3}{5} + \frac{1}{5 \cdot 2} = -\frac{1}{2}, \quad d_1 = -1$$

$$-\frac{\delta_0}{\lambda_0} + \frac{1}{n_0 n_1} + \frac{1}{n_1 n_2} = -\frac{3}{5} + \frac{1}{5 \cdot 2} + \frac{1}{2 \cdot 3} = -\frac{1}{3}, \quad d_2 = -1$$

$$-\frac{\delta_0}{\lambda_0} + \frac{1}{n_0 n_1} + \frac{1}{n_1 n_2} + \frac{1}{n_2 n_3} = -\frac{3}{5} + \frac{1}{5 \cdot 2} + \frac{1}{2 \cdot 3} + \frac{1}{3 \cdot 4} = -\frac{1}{4},$$

$$\left(\equiv \frac{\delta_1}{\lambda_1} \pmod 1 \right) \; d_3 = -1.$$

In the general case, we have essentially the same picture. To get

$$f' : A_m \to A_1,$$

we first define

$$f' f^{m-1}$$

on the white annuli to be the prescribed rotations (Condition (ii)). Then the screw numbers in the black annuli are automatically adjusted as prescribed by Condition (iii), thanks to Claim (F). (Again we must avoid artificial full twists.) On each black annulus this twist is rel.∂ isotopic to a linear twist (Theorem 2.3 and Lemma 2.1). Thus we have obtained the desired

$$f' : A_m \to A_1.$$

Define

$$f' : \mathscr{A} \to \mathscr{A}$$

by setting

$$f'|A_\alpha = f|A_\alpha, \quad \alpha = 1,\ 2,\ \ldots,\ m-1,$$
$$f'|A_m = \text{the above } f'.$$

Then

$$f : \mathscr{A} \to \mathscr{A}$$

is *rel.*∂ isotopic to

$$f' : \mathscr{A} \to \mathscr{A}$$

and in each A_α,

$$(f')^m : A_\alpha \to A_\alpha$$

behaves just as described by Conditions (i), (ii), (iii) and (iv).

This completes the *re-normalization* of

$$f : \mathscr{A} \to \mathscr{A},$$

and we say that the re-normalized

$$f' : \mathscr{A} \to \mathscr{A}$$

is in *superstandard form.*

We will next construct a chorizo space $Ch(\mathscr{A})$ and a pinched covering

$$\pi : \mathscr{A} \to Ch(\mathscr{A})$$

so that $\pi : \mathscr{A} \to Ch(\mathscr{A})$ is a generalized quotient of the renormalized f'.

Take two disks D_0, D_l and $l-1$ spheres

$$S_1,\ S_2,\ \ldots,\ S_{l-1}.$$

Let q_0 and q_1 be the centers of D_0 and D_l, respectively, and let q_i^+ and q_i^- be the north and the south poles of S_i as before. The chorizo space $Ch(\mathscr{A})$ is constructed as shown in Fig. 3.4 (though not yet numerical).

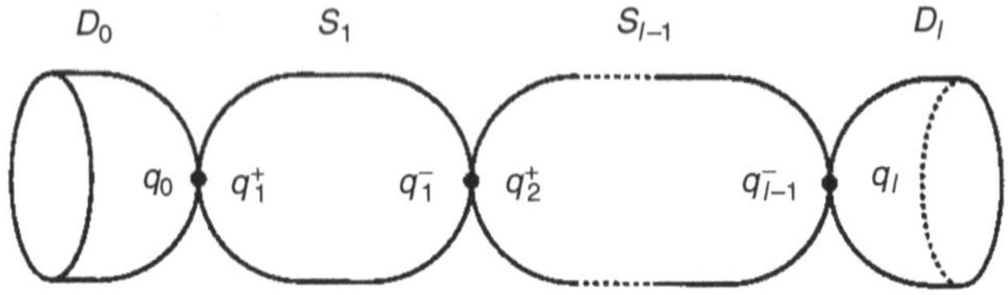

Fig. 3.4 Chorizo space $Ch(\mathscr{A})$

The construction of

$$\pi : \mathscr{A} \to Ch(\mathscr{A})$$

is almost the same as that of

$$\pi : \mathscr{D} \to Ch(\mathscr{D}).$$

First we define

$$\pi : A_1 \to Ch(\mathscr{A}).$$

For this, take the center-line C_k of a black annulus

$$Z_{2k-1}, \quad k = 1, 2, \ldots, l$$

and define $\pi(C_k)$ to be the node

$$q_k^+ \in Ch(\mathscr{A}).$$

The complement $Z_{2k-1} - C_k$ consists of two half-open annuli

$$Z'_{2k-1}, \quad Z''_{2k-1}$$

(Z'_{2k-1} denoting the lower one with respect to the t-level). Define

$$\pi | Z''_{2k-1} \cup Z_{2k} \cup Z'_{2k+1} : Z''_{2k-1} \cup Z_{2k} \cup Z'_{2k+1} \to S_k - \{q_k^-, q_k^+\}$$

to be the n_k-fold cyclic covering whose covering translations are generated by the rotation $(t, x) \to (t, x + 1/n_k)$ (in the (t, x)-picture), $k = 1, \ldots, l-1$. Finally, define

$$\pi | Z'_1 : Z'_1 \to D_0 - \{q_0\}$$

and

$$\pi | Z''_{2l-1} : Z''_{2l-1} \to D_l - \{q_l\}$$

to be the obvious $n_0 (= \lambda_0)$-fold and $n_l (= \lambda_1)$-fold cyclic coverings, respectively.

The construction of

$$\pi | A_\alpha : A_\alpha \to Ch(\mathscr{A})$$

for the other annuli

$$A_\alpha \quad \alpha = 2, \ \ldots, \ m$$

proceeds similarly to that of

$$\pi|\Delta_\alpha : \Delta_\alpha \to Ch(\mathscr{D})$$

for $\alpha = 2, \ \ldots, \ m$.

We repeat it for completeness. First we parametrize the two banks of

$$\overline{N}_k := \pi(Z_{2k-1}),$$

for each $k = 1, \ldots, l$. Let the banks $\pi(Z'_{2k-1})$ and $\pi(Z''_{2k-1})$ be denoted by D^-_{k-1} and D^+_k, respectively. We parametrize them so that if we identify D^-_{k-1} and D^+_k with

$$\{z \mid |z| \le 1\}$$

through the parametrizations,

$$\pi|A_1 : Z'_{2k-1} \cup C_k \to D^-_{k-1}$$

is described as

$$\pi(t, \, x) = \frac{b-t}{b-a} \exp(\sqrt{-1}\, 2\pi \; n_{k-1} \; x)$$

and

$$\pi|A_1 : Z''_{2k-1} \cup C_k \to D^+_k$$

is described as

$$\pi(t, \, x) = \frac{t-b}{c-b} \exp(-\sqrt{-1}\, 2\pi \; n_k \; x),$$

where the annuli

$$Z'_{2k-1} \cup C_k$$

and

$$Z''_{2k-1} \cup C_k$$

are regarded, for simplicity, as

$$Z'_{2k-1} \cup C_k = \{(t, \, x) |\; a \le t \le b, \; x \in \mathbf{R}/\mathbf{Z}\},$$
$$Z''_{2k-1} \cup C_k = \{(t, \, x) |\; b \le t \le c, \; x \in \mathbf{R}/\mathbf{Z}\}.$$

Note that with these parametrizations of D^-_{k-1}, D^+_k, the following identities hold:

$$(T^-_{k-1})^m (\pi|A_1) = (\pi|A_1)(f')^m \quad \text{on} \quad Z'_{2k-1} \cup C_k$$

and

$$(T_k^+)^m(\pi|A_1) = (\pi|A_1)(f')^m \quad \text{on} \quad Z''_{2k-1} \cup C_k,$$

where

$$T_{k-1}^- : D_{k-1}^- \to D_{k-1}^-$$

and

$$T_k^+ : D_k^+ \to D_k^+$$

are defined by

$$T_{k-1}^-(z) = z\exp\left(\frac{\sqrt{-1}\pi}{m\,n_k}(1-|z|)\right)$$

and

$$T_k^+(z) = z\exp\left(\frac{\sqrt{-1}\pi}{m\,n_{k-1}}(1-|z|)\right),$$

respectively, just as in the case of $Ch(\mathscr{D})$.

The construction of

$$\pi|A_\alpha : A_\alpha \to Ch(\mathscr{A})$$

for $\alpha = 2, \ldots, m$ is as follows:

$$\pi|A_\alpha = \begin{cases} (T_{k-1}^-)^{\alpha-1}(\pi|A_1)(f^{\alpha-1})^{-1} & \text{on} \quad f^{\alpha-1}(Z'_{2k-1} \cup C_k), \quad k = 1, \cdots, l, \\ (T_k^+)^{\alpha-1}(\pi|A_1)(f^{\alpha-1})^{-1} & \text{on} \quad f^{\alpha-1}(Z''_{2k-1} \cup C_k), \quad k = 1, \cdots, l, \\ (\pi|A_1)(f^{\alpha-1})^{-1} & \text{on} \quad f^{\alpha-1}(Z_{2k}), \qquad k = 1, \cdots, l. \end{cases}$$

Taking the disjoint union, we get the pinched covering

$$\pi = \bigcup_{\alpha=1}^{m}(\pi|A_\alpha) : \mathscr{A} \to Ch(\mathscr{A}).$$

The images

$$\overline{N}_k := \pi(Z_{2k-1}), \quad k = 1, \ldots, l$$

make the system of closed nodal neighborhoods of $q_1^+, \ldots, q_l^+$ which should be fixed in the definition of a generalized quotient. To make $Ch(\mathscr{A})$ numerical, attach the multiplicities

$$m\,n_0,\ m\,n_1,\ \ldots,\ m\,n_l$$

to the components

$$D_0,\ S_1,\ \ldots,\ S_{l-1},\ D_l.$$

With the system $\{N_k\}_{k=1}^{l}$ and the multiplicities, one can easily check Conditions (i) $\sim$ (v) (of Def. 3.4) to be satisfied by a generalized quotient of

$$f' : \mathscr{A} \to \mathscr{A}.$$

We summarize the above argument.

Lemma 3.3. *Let $\mathscr{A}$ be a disjoint union of m annuli,*

$$A_1,\ A_2,\ \ldots,\ A_m.$$

Suppose that a homeomorphism

$$f : \mathscr{A} \to \mathscr{A}$$

permutes these annuli cyclically, and that for each α $(= 1, 2, \ldots, m)$,

$$f^m : A_\alpha \to A_\alpha$$

is a linear twist of negative screw number s. Let

$$(m,\ \lambda_0,\ \sigma_0)$$

and

$$(m,\ \lambda_1,\ \sigma_1)$$

be the valencies of the two boundary curves of A_α (which are independent of α).

Let

$$(n_0,\ n_1,\ \ldots,\ n_l),\quad l \geq 1,$$

be the sequence of positive integers which are determined by

$$n_0 = \lambda_0,\quad n_l = \lambda_1,$$
$$n_1 \equiv \sigma_0 \pmod{\lambda_0},\quad n_{l-1} \equiv \sigma_1 \pmod{\lambda_1},$$
$$n_{i-1} + n_{i+1} \equiv 0 \pmod{n_i},\quad i = 1,\ 2,\ \ldots,\ l-1,$$
$$\frac{(n_{i-1} + n_{i+1})}{n_i} \geq 2,\quad i = 1,\ 2,\ \ldots\ l-1,$$

and

$$\sum_{i=0}^{l-1} \frac{1}{n_i\, n_{i+1}} = |s|.$$

(For the existence and the uniqueness of such a sequence, see Theorem 5.1.)

Let $Ch(\mathscr{A})$ be the chorizo space as shown in Fig. 3.4, *and attach the multiplicities*

$$m\,n_0,\ m\,n_1,\ \ldots,\ m\,n_l,$$

to the components

$$D_0,\ S_1,\ \ldots,\ S_{l-1},\ D_l.$$

Then there exist a pinched covering

$$\pi : \mathscr{A} \to Ch(\mathscr{A})$$

and a rel.∂ isotopy from

$$f : \mathscr{A} \to \mathscr{A}$$

to

$$f' : \mathscr{A} \to \mathscr{A}$$

such that

$$\pi : \mathscr{A} \to Ch(\mathscr{A})$$

is a generalized quotient of f'.

Remark 3.5. By the property of

$$(n_0,\ n_1,\ \ldots,\ n_l),$$

the sphere components

$$S_1,\ S_2,\ \ldots,\ S_{l-1}$$

in $Ch(\mathscr{A})$ satisfy the minimality condition.

In the general case when

$$f : \mathscr{A} \to \mathscr{A}$$

does not necessarily permute the annuli cyclically, decompose the annuli into cyclic orbits:

$$\mathscr{A} \to \mathscr{A}^{(1)} \cup \mathscr{A}^{(2)} \cup \ldots \cup \mathscr{A}^{(s)},$$

and define

$$\pi : \mathscr{A} \to Ch(\mathscr{A})$$

to be the disjoint union

$$\bigcup_{\nu=1}^{s} \mathscr{A}^{(\nu)} \to \bigcup_{\nu=1}^{s} Ch(\mathscr{A}^{(\nu)}).$$

3.6 Re-normalization of a Special Twist

Let

$$\mathscr{A} = \bigcup_{\alpha=1}^{m} A_\alpha$$

be a disjoint union of annuli. Suppose that a homeomorphism

$$f : \mathscr{A} \to \mathscr{A}$$

cyclically permutes the components, and that for each $\alpha = 1, 2, \ldots, m$,

$$f^m : A_\alpha \to A_\alpha$$

is a special twist (in the sense of Chap. 2) of *negative* screw number. By the definition of a special twist, every annulus in $\mathscr{A}$ is amphidrome. To distinguish the present case from the previous (linear) case, we will denote the disjoint union of annuli by $\mathscr{A}_{\mathrm{sp}}$ in the case of specials twist, and by $\mathscr{A}_{\mathrm{ln}}$ in the case of linear twists.

We may assume

$$f(A_\alpha) = A_{\alpha+1}, \quad \alpha = 1, \ldots, m-1, \quad f(A_m) = A_1$$

as before.

We will construct a chorizo space $Ch(\mathscr{A}_{\mathrm{sp}})$ and a pinched covering

$$\pi : \mathscr{A}_{\mathrm{sp}} \to Ch(\mathscr{A}_{\mathrm{sp}}),$$

and will show that

$$f : \mathscr{A}_{\mathrm{sp}} \to \mathscr{A}_{\mathrm{sp}}$$

is *rel.*∂ isotopic to a homeomorphism

$$f' : \mathscr{A}_{\mathrm{sp}} \to \mathscr{A}_{\mathrm{sp}}$$

such that

$$\pi : \mathscr{A}_{\mathrm{sp}} \to Ch(\mathscr{A}_{\mathrm{sp}})$$

is a generalized quotient of f'. This, however, has been essentially done by the compound nature of a special twist.

Let us identify A_1 with

$$[0, 1] \times S^1,$$

where S^1 is $\mathbf{R}/\mathbf{Z}$. Let $s(<0)$ be the screw number of f in A_1. Then by Corollary 2.5, we have that

$$f^m(t,x) = \begin{cases} \left(1-t,\ -x + \frac{3}{4}s\left(t-\frac{1}{3}\right)\right) & \text{for } 0 \le t \le \frac{1}{3}, \\ (1-t,\ -x) & \text{for } \frac{1}{3} \le t \le \frac{2}{3}, \\ \left(1-t,\ -x + \frac{3}{4}s\left(t-\frac{2}{3}\right)\right) & \text{for } \frac{2}{3} \le t \le 1. \end{cases}$$

We decompose

$$A_1 = [0,1] \times S^1$$

into three parts:

$$\left[0, \frac{1}{3}\right] \times S^1, \quad \left[\frac{1}{3}, \frac{2}{3}\right] \times S^1, \quad \left[\frac{2}{3}, 1\right] \times S^1.$$

Let A'_1 denote

$$\left[0, \frac{1}{3}\right] \times S^1 \subset A_1,$$

and for each $\alpha = 1, 2, \ldots, 2m$, set

$$A'_\alpha = f^{\alpha-1}(A'_1).$$

Clearly

$$A'_{m+1} = \left[\frac{2}{3}, 1\right] \times S^1 \subset A_1,$$

and more generally

$$A'_\alpha \cup A'_{\alpha+m} \subset A_\alpha, \quad \alpha = 1, 2, \ldots, m.$$

Let $\mathscr{A}'_{\text{ln}}$ denote the disjoint union

$$\bigcup_{\alpha=1}^{2m} A'_\alpha.$$

If

$$f : \mathscr{A}_{\text{sp}} \to \mathscr{A}_{\text{sp}}$$

is restricted to

$$f : \mathscr{A}'_{\text{ln}} \to \mathscr{A}'_{\text{ln}},$$

it cyclically permutes the $2m$ annuli.

Claim (G). The map

$$f^{2m} : A'_1 \to A'_1$$

is a linear twist with screw number $1/2s$.

Proof. Using the above expression of f^m, we compute f^{2m} to get

$$f^{2m}(t, x) = \left(t, x - \frac{3}{2}st + \frac{1}{2}s\right), \quad 0 \le t \le \frac{1}{3}.$$

Since A'_1 is parametrized as

$$\left[0, \frac{1}{3}\right] \times S^1,$$

the screw number is the coefficient of $-3t$, which is $1/2s$. (Cf. Corollary 2.2). □

Let

$$(2m, \lambda_0, \sigma_0)$$

be the valency of the "outer" boundary $\partial_0 A'_1 = \{0\} \times S^1$ with respect to

$$f : \mathscr{A}'_{\mathrm{In}} \to \mathscr{A}'_{\mathrm{In}}.$$

Let δ_0 be the integer determined by

$$\sigma_0\delta_0 \equiv 1 \pmod{\lambda_0}, \quad 0 \le \delta_0 \le \lambda_0.$$

(NB. $\delta_0 = 0$ iff $\lambda_0 = 1$). Then

$$\delta_0/\lambda_0 \equiv -\frac{1}{2}s \pmod 1,$$

see Corollary 2.3. The valency of the "inner" boundary $\partial_{1/3} A'_1 = \{1/3\} \times S^1$ is equal to

$$(2m, 1, 0).$$

(This is geometrically obvious, but also can be seen from the expression of $f^{2m}(t, x)$ in the proof of Claim (G).)

Applying Claim (F) (1), we obtain a unique

$$(n_0, n_1, \ldots, n_l)$$

of positive integers such that

(α') $n_0 = \lambda_0, \quad n_l = 1$,
(β') $n_1 \equiv \sigma_0 \pmod{\lambda_0}$,
(γ') $n_{i-1} + n_{i+1} \equiv 0 \pmod{n_i}, \quad i = 1, 2, \ldots, l-1$,
(δ') $(n_{i-1} + n_{i+1})/n_i \ge 2, \quad i = 1, 2, \ldots, l-1$, and
(ε') $\sum_{i=0}^{l-1} 1/n_i n_{i+1} = 1/2|s|$.

In fact, in the present case, the integers

$$(n_0, n_1, \ldots, n_l)$$

can be determined more easily.

Case 1. $\delta_0 = 0$; then $\lambda_0 = 1$, and

$$\frac{1}{2}\,|s|$$

is an integer $k (> 0)$ because

$$\frac{\delta_0}{\lambda_0} \equiv -\frac{1}{2}\,s \pmod 1.$$

The sequence is nothing but a sequence of $k + 1$ copies of 1's:

$$(\underbrace{1,\ 1,\ \ldots,\ 1}_{k+1}).$$

Case 2. $\delta_0 > 0$; determine $(n_0, n_1, \ldots, n_k)$ using the Euclidean algorithm, so that

$$n_0 = \lambda,\ n_1 = \sigma, \quad n_0 > n_1 > \cdots > n_k = 1,$$

$$n_{i-1} + n_{i+1} \equiv 0 \pmod{n_i}, \quad i = 1,\ \ldots,\ k-1.$$

Then by Lemma 5.2(3) (in Chap. 5), we have

$$\sum_{i=0}^{k-1} \frac{1}{n_i\, n_{i+1}} = \frac{\delta_0}{\lambda_0}.$$

Since

$$\frac{1}{2}\,|s| - \frac{\delta_0}{\lambda_0} \;=\; \text{an integer (say } h) \geq 0,$$

we can add h copies of 1's to obtain

$$(n_0,\ n_1,\ \ldots,\ n_l) \;=\; (n_0,\ \ldots,\ n_k,\ \underbrace{1,\ 1,\ \ldots,\ 1}_{h}).$$

In this case $l = k + h$.

Remark 3.6. In both cases,

$$(n_0,\ n_1,\ \ldots,\ n_l)$$

is in the form

$$n_0 \;>\; n_1 \;>\; \cdots \;> n_k \;=\; n_{k+1} \;=\; \cdots \;=\; n_l \;=\; 1,$$

for some k with $0 \leq k \leq l$.

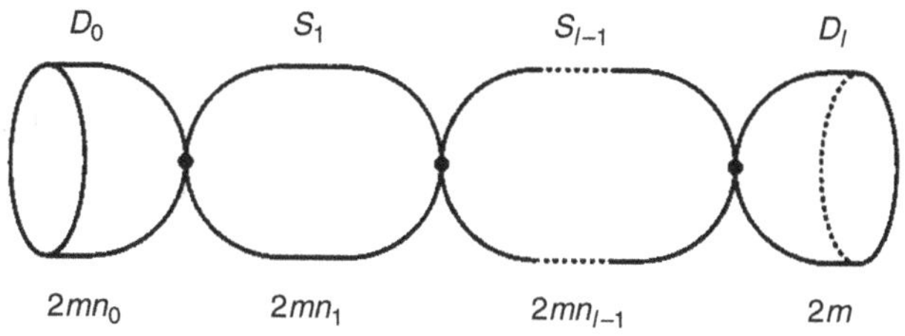

Fig. 3.5 Chorizo space $Ch(\mathscr{A}'_{ln})$

By Lemma 3.3, we have a generalized quotient

$$\pi : \mathscr{A}'_{\mathrm{ln}} \to Ch(\mathscr{A}'_{\mathrm{ln}})$$

of the re-normalized

$$f' : \mathscr{A}'_{\mathrm{ln}} \to \mathscr{A}'_{ln}.$$

The shape of $Ch(\mathscr{A}'_{\mathrm{ln}})$ is shown in Fig. 3.5.

We have next to consider the middle annulus

$$B'_1 = \left[\frac{1}{3}, \frac{2}{3}\right] \times S^1 \subset A_1.$$

For $\alpha = 1, \ldots, m$, we set

$$B'_\alpha = f^{\alpha-1}(B'_1),$$

and let $\mathscr{B}'$ denote the disjoint union

$$\bigcup_{\alpha=1}^{m} B'_\alpha.$$

The m-th iteration of f restricted to B'_α gives a "180°-rotation"

$$f^m : B'_\alpha \to B'_\alpha$$

around an axis. See Fig. 3.6.

Let p_1^α, p_2^α be the fixed points of

$$f^m : B'_\alpha \to B'_\alpha$$

and let

$$D_\nu^\alpha, \quad \nu = 1, 2$$

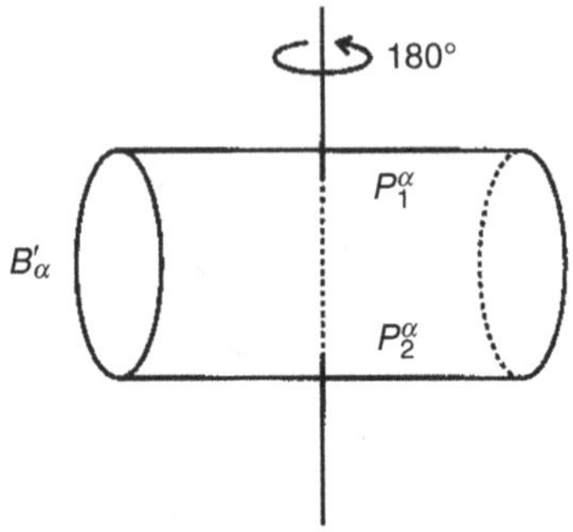

Fig. 3.6 $f^m : B'_\alpha \to B'_\alpha$ is a "180°-rotation"

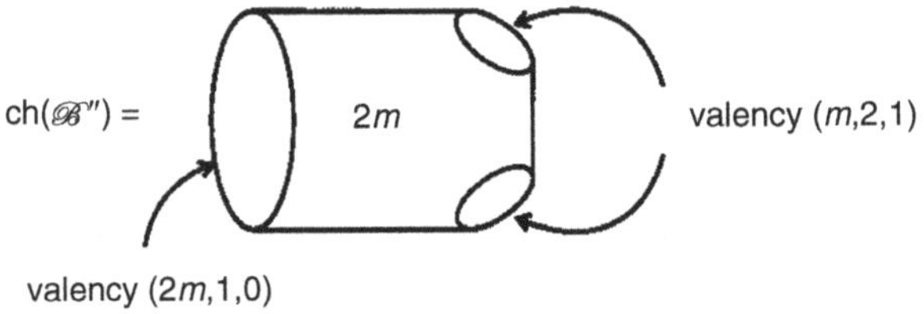

Fig. 3.7 Chorizo space $Ch(\mathscr{B}'')$

be a small invariant disk neighborhood of p_ν^α. We denote the disjoint union

$$\bigcup_{\alpha=1}^{m} D_\nu^\alpha$$

by $\mathscr{D}_\nu$. The complement

$$\mathscr{B}'' = \mathscr{B}' - (Int(\mathscr{D}_1) \cup Int(\mathscr{D}_2))$$

has no multiple point, and f acts on $\mathscr{B}''$ as a periodic map of order $2m$. We regard the quotient $\mathscr{B}''/f$ as a chorizo space (without nodes) and denote it by $Ch(\mathscr{B}'')$, whose shape is shown in Fig. 3.7.

As for the union of m disks $\mathscr{D}_\nu$, Lemma 3.2 can be applied to produce a chorizo space $Ch(\mathscr{D}_\nu)$. See Fig. 3.8.

To obtain a generalized quotient corresponding to

$$f : \mathscr{A}_{\rm sp} \to \mathscr{A}_{\rm sp},$$

we have only to paste together the chorizo spaces

$$Ch(\mathscr{A}'_{\rm ln}), \quad Ch(\mathscr{B}''), \quad Ch(\mathscr{D}_\nu), \qquad \nu = 1,\ 2,$$

along their boundaries. The multiplicities match up naturally. The shape of $Ch(\mathscr{A}_{\rm sp})$ is shown in Fig. 3.9. It consists of a disks and $l + 2$ spheres.

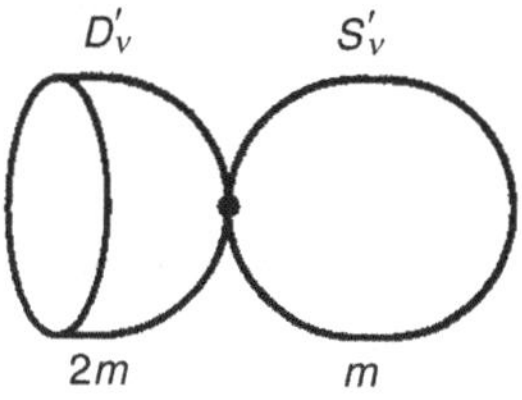

Fig. 3.8 Chorizo space $Ch(\mathscr{D}_\nu)$, $\nu = 1, 2$

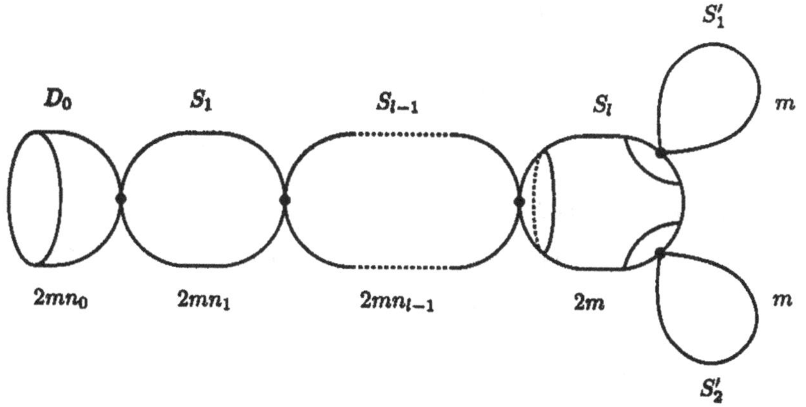

Fig. 3.9 Chorizo space $Ch(\mathscr{A}_{sp})$

The pinched covering

$$\pi : \mathscr{A}_{\text{sp}} \to Ch(\mathscr{A}_{\text{sp}})$$

is the union of

$$\pi : \mathscr{A}'_{\text{ln}} \to Ch(\mathscr{A}'_{\text{ln}}),$$
$$\pi : \mathscr{B}'' \to Ch(\mathscr{B}'')$$

and

$$\pi : \mathscr{D}_\nu \to Ch(\mathscr{D}_\nu), \quad \nu = 1,\ 2.$$

The homeomorphism

$$f : \mathscr{A}_{\text{sp}} \to \mathscr{A}_{\text{sp}}$$

can be re-normalized over $Ch(\mathscr{A}'_{\text{ln}})$, and $Ch(\mathscr{D}_\nu)$, $\nu = 1, 2$, without being changed over $Ch(\mathscr{B}'')$.

We summarize the above argument.

Lemma 3.4. *Let $\mathscr{A}_{\text{sp}}$ be a disjoint union of m annuli,*

$$A_1,\ A_2,\ \ldots,\ A_m.$$

Suppose that a homeomorphism

$$f : \mathscr{A}_{\text{sp}} \to \mathscr{A}_{\text{sp}}$$

permutes these annuli cyclically, and that for each $\alpha = 1, 2, \ldots, m$, *the map*

$$f^m : A_\alpha \to A_\alpha$$

is a special twist of negative screw number s. Let

$$(2m, \lambda, \sigma)$$

be the valency common to the two boundary curves of A_α (which is independent of α). *Let*

$$(n_0, n_1, \ldots, n_l), \quad l \geq 1,$$

be the sequence of positive integers which are uniquely determined by

$$n_0 = \lambda, \quad n_1 \equiv \sigma \pmod{\lambda},$$
$$n_0 > n_1 > \cdots > n_k = n_{k+1} = \cdots = n_l = 1, \quad \textit{for some } k \textit{ with } 0 \leq k \leq l,$$
$$n_{i-1} + n_{i+1} \equiv 0 \pmod{n_i}, \quad i = 1, 2, \ldots, l-1, \quad \textit{and}$$
$$\sum_{i=0}^{k-1} \frac{1}{n_i n_{i+1}} = \frac{1}{2}|s|.$$

Let $Ch(\mathscr{A}_{\mathrm{sp}})$ *be the chorizo space as shown in* Fig. 3.9, *and attach the multiplicities*

$$2mn_0, 2mn_1, \ldots, 2mn_l \; (= 2m), m, m,$$

to the components

$$D_0, S_1, \ldots, S_l, S'_1, S'_2.$$

Then there exist a pinched covering

$$\pi : \mathscr{A}_{\mathrm{sp}} \to Ch(\mathscr{A}_{\mathrm{sp}})$$

and a rel.∂ isotopy from

$$f : \mathscr{A}_{\mathrm{sp}} \to \mathscr{A}_{\mathrm{sp}}$$

to

$$f' : \mathscr{A}_{\mathrm{sp}} \to \mathscr{A}_{\mathrm{sp}}$$

such that

$$\pi : \mathscr{A}_{\mathrm{sp}} \to Ch(\mathscr{A}_{\mathrm{sp}})$$

is a generalized quotient of f'.

Note that the sphere components of $Ch(\mathscr{A}_{\mathrm{sp}})$ satisfy the minimality condition.

In the general case when $f : \mathscr{A}_{\text{sp}} \to \mathscr{A}_{\text{sp}}$ does not necessarily permute the annuli cyclically, decompose the annuli into cyclic orbits as before:

$$\mathscr{A}_{\text{sp}} = A_{\text{sp}}^{(1)} \cup \cdots \cup \mathscr{A}_{\text{sp}}^{(s)},$$

and define

$$\pi : \mathscr{A}_{\text{sp}} \to Ch(\mathscr{A}_{\text{sp}})$$

to be the disjoint union

$$\bigcup_{\nu=1}^{s} \mathscr{A}_{\text{sp}}^{(\nu)} \to \bigcup_{\nu=1}^{s} Ch(\mathscr{A}_{\text{sp}}^{(\nu)}).$$

3.7 Completion of Theorem 3.1. (Existence)

Remember that we had a pseudo-periodic map

$$f : \Sigma_g \to \Sigma_g$$

(of negative twist) in standard form and we tried to isotop f in order that the resulting f' would have its minimal quotient.

At the beginning of the proof, we took a system of annular neighborhoods $\mathscr{A}$ of a precise cut system, on which linear twists or special twists, or both, take place. Outside $\mathscr{A}$, f is a periodic map. Let $\mathscr{B}$ denote the outside:

$$\mathscr{B} = \Sigma_g - Int(\mathscr{A}).$$

We also took a disjoint union $\mathscr{D}$ of invariant disk neighborhoods of all the multiple points of

$$f|\mathscr{B} : \mathscr{B} \to \mathscr{B},$$

and set

$$\mathscr{B}' = \mathscr{B} - Int(\mathscr{D}).$$

We further classified the annuli in $\mathscr{A}$ into $\mathscr{A}_{\text{sp}}$ and $\mathscr{A}_{\text{ln}}$ according to their character of being amphidrome or non-amphidrome.

Now, on all these spaces,

$$\mathscr{B}', \quad \mathscr{D}, \quad \mathscr{A}_{\text{ln}}, \quad \text{and} \quad \mathscr{A}_{\text{sp}},$$

f is *rel.*∂ isotopic to f' in superstandard form ($f'|\mathscr{B}'$ is the same as $f|\mathscr{B}'$), and f', when restricted to one of these spaces, has a generalized quotient

$$\pi : \mathscr{B}' \to Ch(\mathscr{B}'),$$

$$\pi : \mathscr{D} \to Ch(\mathscr{D}),$$

$$\pi : \mathscr{A}_{\mathrm{ln}} \to Ch(\mathscr{A}_{\mathrm{ln}}),$$

or

$$\pi : \mathscr{A}_{\mathrm{sp}} \to Ch(\mathscr{A}_{\mathrm{sp}}).$$

We paste together these chorizo spaces along their boundaries. The multiplicities match up, and we obtain a chorizo space

$$S = Ch(\mathscr{B}') \cup Ch(\mathscr{D}) \cup Ch(\mathscr{A}_{\mathrm{ln}}) \cup Ch(\mathscr{A}_{\mathrm{sp}})$$

and a pinched covering $\pi : \Sigma_g \to S$, which is a generalized quotient of f'.

It remains to show that the sphere components in S satisfy the minimality condition. We have already checked it for spheres in

$$Ch(\mathscr{D}), \quad Ch(\mathscr{A}_{\mathrm{ln}}), \quad \text{and} \quad Ch(\mathscr{A}_{\mathrm{sp}}),$$

so let Θ be a sphere component (without self-intersection) which is not contained in

$$Ch(\mathscr{D}) \cup Ch(\mathscr{A}_{\mathrm{ln}}) \cup Ch(\mathscr{A}_{\mathrm{sp}}).$$

Then the closed part P in Θ is a connected component of $Ch(\mathscr{B}')$. This means that each connected component of $\pi^{-1}(P)$ is a component of $\mathscr{B}'$ which has *negative* Euler characteristic because

$$\mathscr{B}' = \mathscr{B} - Int(\mathscr{D})$$

and $\mathscr{B}$ is the complement of the annular neighborhoods $\mathscr{A}$ of a precise cut system. Therefore

$$\chi(P) < 0,$$

which implies that the sphere component Θ intersects the other irreducible components in more than two points. Thus the minimality condition is trivially satisfied by Θ. This completes the proof of Theorem 3.1. □

Chapter 4
Uniqueness of Minimal Quotient

4.1 Main Theorem of Chap. 4

Definition 4.1. A homeomorphism

$$H : S \to S'$$

between chorizo spaces is *numerical* if it preserves the orientation and the multiplicity on each part.

In this chapter, we will prove the following:

Theorem 4.1 (Uniqueness). *Let f and $f' : \Sigma_g \to \Sigma_g$ be pseudo-periodic maps of negative twist. Suppose there are minimal quotients*

$$\pi : \Sigma_g \to S$$

and

$$\pi' : \Sigma_g \to S'$$

of f and f' respectively (thus f and f' are in superstandard form). If f and f' are mutually homotopic, then there exist a homeomorphism

$$h : \Sigma_g \to \Sigma_g$$

and a numerical homeomorphism

$$H : S \to S'$$

such that

Y. Matsumoto and J.M. Montesinos-Amilibia, *Pseudo-periodic Maps and Degeneration of Riemann Surfaces*, Lecture Notes in Mathematics 2030,
DOI 10.1007/978-3-642-22534-5_4,

(i) h is isotopic to the identity,
(ii) $f = h^{-1} f' h$, *and*
(iii) $\pi' h = H \pi$:

$$\begin{array}{ccc} \Sigma_g & \xrightarrow{h} & \Sigma_g \\ \pi \downarrow & & \downarrow \pi' \\ S & \xrightarrow[H]{} & S' \end{array}$$

We may assume that one of these quotients, say,

$$\pi' : \Sigma_g \to S'$$

is constructed by the method of Chap. 3 as the union of

$$Ch(\mathscr{B}'), \; Ch(\mathscr{D}), \; Ch(\mathscr{A}_{\text{ln}}) \text{ and } Ch(\mathscr{A}_{\text{sp}}).$$

We will study the structure of the other one $\pi : \Sigma_g \to S$.

Let us start with a set of general definitions.

Definition 4.2. An irreducible component Θ of a chorizo space S is said to be *trivial* if Θ is a sphere without self-intersection and intersects the union of the other irreducible components in one or two points. Otherwise Θ is said to be *essential.*

Definition 4.3. Let Θ be a trivial component. Assume there exist a sequence

$$\Theta = \Theta_0, \; \Theta_1, \; \ldots, \; \Theta_m$$

of trivial components such that Θ_i intersects $\Theta_{i+1}, i = 0, \ldots, m-1$, and the last Θ_m intersects only Θ_{m-1} in one point exactly. Then Θ is called a trivial component of *type I* (of course, all members in the sequence are trivial components of type I, see Fig. 4.1). A trivial component which does not touch any trivial component of type I is said to be of *type II.*

Definition 4.4. Let T_1 (resp. T_2) be the subspace of S which is the union of all trivial components of type I (resp. type II). A connected component of T_1 (resp. T_2) will be called a *trivial array of type I* (*resp. type II*).

Hereafter, a system of closed nodal neighborhoods

$$\{\overline{N}_p\}_{p\,=\,\text{node}}$$

will be fixed in S (as in the case of the "base chorizo" of a generalized quotient. See Remark 1) before Lemma 3.1).

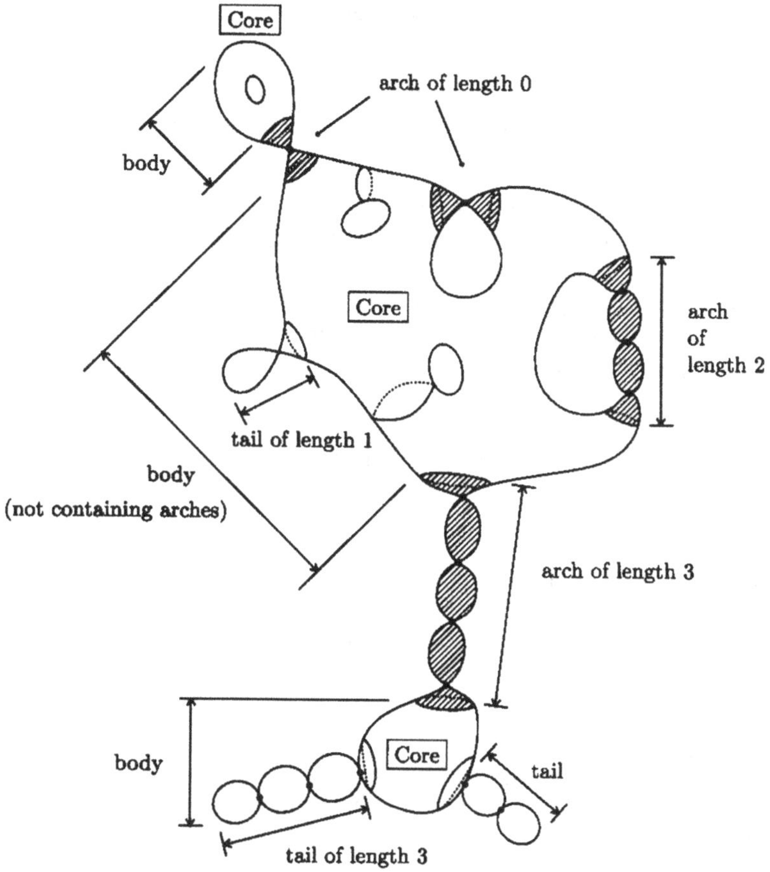

Fig. 4.1 Decomposition of a chorizo space

As before, from now on, when we talk about a closed nodal neighborhood $\overline{N}_p$ and a closed part P, we always will be talking about a member of the fixed system

$$\{\overline{N}_p\}_{p\,=\,\text{node}}$$

and a connected component of

$$S - \bigcup_{p\,=\,\text{node}} N_p,$$

respectively, where

$$N_p = Int(\overline{N}_p).$$

Definition 4.5. Any union of closed nodal neighborhoods and closed parts is called a *subchorizo space* .

Definition 4.6. A *tail* of S is a minimal subchorizo space containing a given trivial array of type I (see Fig. 4.1). The number of the trivial components contained in the array is the *length* of the tail, which is always ≥ 1.

Definition 4.7. An *arch* of S is a minimal subchorizo space containing a given trivial array of type II. The number of the trivial components contained in the array is the *length* of the arch, which is ≥ 1. In general, there still remain some nodes which are not in tails nor in arches. The closed nodal neighborhood of such a node is called an *arch of length* 0. (Such a node is either a self- intersection point of an essential component or an intersection point between two essential components.)

Let

$$ARCH = \bigcup_{\nu=1}^{s} ARCH_\nu$$

denote the union of all arches (of length ≥ 0) in S.

Definition 4.8. A connected component of

$$S - Int(ARCH)$$

is called a *body*.

Here is an elementary lemma.

Lemma 4.1. *A body contains one and only one closed part with negative Euler characteristic.*

Proof. Note that there is a bijective correspondence between closed parts and irreducible components: $P \leftrightarrow \Theta$ (such that $P \subset \Theta$), and that a closed part P has negative Euler characteristic if and only if the corresponding irreducible component Θ is an essential component. (A "pinched torus" is the only example of an essential component that contains a closed part of Euler characteristic 0. But this does not appear in the "base chorizo space" of a surface Σ_g with genus $g \geq 2$.)

Suppose a body BDY_ν (ν being a suffix) contains a closed part P whose Euler characteristic is ≥ 0. Then P belongs to a trivial component of type I (because trivial components of type II have been deleted), thus it belongs to a tail. But the tail is attached to an essential component Θ_μ (μ being a suffix), and the closed part P_μ of Θ_μ belongs to the BDY_ν. This proves that BDY_ν contains a closed part P_μ with $\chi(P_\mu) < 0$.

Since two essential components are joined only by arches, two closed parts in different essential components cannot belong to the same connected component of $S - int(ARCH)$. Thus BDY_ν contains only one closed part with negative Euler characteristic. □

Definition 4.9. The closed part with negative Euler characteristic is called the *core part* of the body.

Remark 4.1. A body is a subchorizo space which is a union of its core part P and the tails that are attached to P. Figure 4.1 shows an example of the decomposition of a chorizo space.

Now we return to the generalized quotient

$$\pi : \Sigma_g \to S$$

of Theorem 4.1. Remember that it is a minimal generalized quotient of a pseudo-periodic map

$$f : \Sigma_g \to \Sigma_g$$

of negative twist. Let us examine the structure of

$$\pi^{-1}(\text{arch}), \quad \pi^{-1}(\text{tail}), \quad \text{and} \quad \pi^{-1}(\text{body}).$$

4.2 Structure of π^{-1}(arch)

Let $ARCH_\nu$ be an arch. To make the notation consistent with that of Lemma 3.3, let $l-1$ denote the length of the arch. $ARCH_\nu$ contains two attaching banks D_0, D_l. If the length is ≥ 1 (i.e. if $l \geq 2$) it also contains $l-1$ spheres $S_1, \ldots, S_{l-1}$ which are trivial components of type II. Suppose

$$D_0 \cap S_1 = \{q_1\}, \quad S_{i-1} \cap S_i = \{q_i\}, \quad i = 2, \ldots, l-1$$

and $S_{l-1} \cap D_l = \{q_l\}$.

Let m_0, m_l (or m_i) denote the multiplicities of D_0, D_l (or S_i), $i = 1, \ldots, l-1$. By Proposition 3.1,

$$m_{i-1} + m_{i+1} \equiv 0 \pmod{m_i}, \quad i = 1, \ldots, l-1.$$

Thus

$$\gcd(m_{i-1}, m_i)$$

is independent of $i (= 1, \ldots, l)$, which will be denoted by m.

Let $\overline{N}_i$ be the closed nodal neighborhood of q_i, $i = 1, \ldots, l$. Then by the definition of a generalized quotient (Chap. 3),

$$\pi^{-1}(\overline{N}_i)$$

consists of m annuli which are permuted by

$$f : \pi^{-1}(\overline{N}_i) \to \pi^{-1}(\overline{N}_i)$$

cyclically.

The closed part P_i contained in the sphere

$$S_i, \quad i = 1, \dots, l-1,$$

is an annulus. Thus

$$\pi^{-1}(P_i)$$

is also a disjoint union of annuli, and the number of the annuli in

$$\pi^{-1}(P_i)$$

is equal to m.

These "narrow" annuli in

$$\pi^{-1}(\overline{N}_i), \quad i = 1, \dots, l,$$

and those in

$$\pi^{-1}(P_i), \quad i = 1, \dots, l-1,$$

are pasted together along their boundaries to make a disjoint union of m, "long" annuli,

$$A_1, A_2, \dots, A_m.$$

Thus

$$\pi^{-1}(ARCH_\nu) = \bigcup_{\alpha=1}^{m} A_\alpha.$$

The map

$$f : \pi^{-1}(ARCH_\nu) \to \pi^{-1}(ARCH_\nu)$$

permutes

$$A_1, A_2, \dots, A_m$$

cyclically.

Divide the unit interval $[0, 1]$ into $2l - 1$ small intervals

$$I_1, I_2, \dots, I_{2l-1},$$

where

$$I_j = \left[\frac{j-1}{2l-1}, \frac{j}{2l-1}\right].$$

Lemma 4.2. *There are parametrizations*

$$\phi_\alpha : [0, 1] \times S^1 \to A_\alpha \ (S^1 = \mathbf{R}/\mathbf{Z}), \quad \alpha = 1, \dots, m,$$

such that

(i) $\pi^{-1}(\overline{N}_i) = \bigcup_{\alpha=1}^{m} \phi_\alpha(I_{2i-1} \times S^1), \quad i = 1, \ldots, l,$

(ii) $\pi^{-1}(P_i) = \bigcup_{\alpha=1}^{m} \phi_\alpha(I_{2i} \times S^1), \quad i = 1, \ldots, l-1,$

(iii) *the map*

$$\phi_\alpha^{-1} f^m \phi_\alpha : I_{2i-1} \times S^1 \to I_{2i-1} \times S^1, \ (\textit{for each } \alpha = 1, \ldots, m)$$

is a linear twist with screw number

$$-\frac{1}{n_{i-1}n_i},$$

where

$$n_{i-1} = \frac{m_{i-1}}{m}, \quad n_i = \frac{m_i}{m}, \quad i = 1, \ldots, l,$$

and

(iv) *the map*

$$\phi_\alpha^{-1} f^m \phi_\alpha : I_{2i} \times S^1 \to I_{2i} \times S^1 \ (\textit{for each } \alpha = 1, \ldots, m)$$

is a rotation sending

$$(t, x) \quad \textit{to} \quad \left(t, x + \frac{d_i}{n_i}\right),$$

where d_i is an integer coprime with n_i, $i = 1, \ldots, l-1$.

(v) *Let*

$$(m, \lambda_0, \sigma_0) \quad \textit{and} \quad (m, \lambda_1, \sigma_1)$$

be the valencies of

$$\partial_0 A_\alpha = \phi_\alpha(\{0\} \times S^1)$$

and

$$\partial_1 A_\alpha = \phi_\alpha(\{1\} \times S^1),$$

respectively. Define the integers δ_0, δ_1 as usual:

$$0 \leq \delta_i < \lambda_i, \ \sigma_i \delta_i \equiv 1 \pmod{\lambda_i}, \quad i = 0, 1.$$

Then

$$f^m \phi_\alpha(0, x) = \phi_\alpha\left(0, x - \frac{\delta_0}{\lambda_0}\right)$$

and

$$f^m \phi_\alpha(1, x) = \phi_\alpha\left(1,\ x + \frac{\delta_1}{\lambda_1}\right).$$

Proof. First we fix our attention at the "first annulus" A_1. Set

$$Y_{2i-1} = \pi^{-1}(\overline{N}_i) \cap A_1 \quad \text{(a "black" annulus)}$$

and

$$Y_{2i} = \pi^{-1}(P_i) \cap A_1 \quad \text{(a "white" annulus).}$$

Of course,

$$A_1 = \bigcup_{j=1}^{2l-1} Y_j.$$

By the definition of a generalized quotient, there is a parametrization of a black annulus

$$\psi_{2i-1} : I_{2i-1} \times S^1 \to Y_{2i-1},$$

for each $i = 1, \dots, l$, such that

$$\psi_{2i-1}^{-1} f^m \psi_{2i-1} : I_{2i-1} \times S^1 \to I_{2i-1} \times S^1$$

is a linear twist with screw number

$$-\frac{1}{n_{i-1} n_i}.$$

We can assume

$$\psi_{2i-1}\left(\left\{\frac{2i-2}{2l-1}\right\} \times S^1\right) = Y_{2i-2} \cap Y_{2i-1},$$

$$\psi_{2i-1}\left(\left\{\frac{2i-1}{2l-1}\right\} \times S^1\right) = Y_{2i-1} \cap Y_{2i}.$$

For a white annulus Y_{2i}, we take a preliminary parametrization

$$\psi'_{2i} : I_{2i} \times S^1 \to Y_{2i},$$

satisfying

$$\psi'_{2i}|_{\{(2i-1)/(2l-1)\}\times S^1} = \psi_{2i-1}|_{\{(2i-1)/(2l-1)\}\times S^1}$$

and

$$\psi'_{2i}|_{\{2i/(2l-1)\}\times S^1} = \psi_{2i+1}|_{\{2i/(2l-1)\}\times S^1},$$

so that ψ'_{2i} is "linear" on the boundary of Y_{2i}. Since

$$\pi : Y_{2i} \to P_i$$

is an n_i-fold cyclic covering over an *annulus* P_i, there is an integer d_i, coprime with n_i, such that on ∂Y_{2i} the following hold:

$$f^m \psi'_{2i}\left(\frac{2i-1}{2l-1}, x\right) = \psi'_{2i}\left(\frac{2i-1}{2l-1}, x + \frac{d_i}{n_i}\right),$$

$$f^m \psi'_{2i}\left(\frac{2i}{2l-1}, x\right) = \psi'_{2i}\left(\frac{2i}{2l-1}, x + \frac{d_i}{n_i}\right).$$

Then the parametrization

$$\psi'_{2i} : I_{2i} \times S^1 \to Y_{2i}$$

can be rectified (by an isotopy *rel.*∂) to a parametrization

$$\psi_{2i} : I_{2i} \times S^1 \to Y_{2i}$$

satisfying condition (iv) of the lemma.

The desired parametrization for A_1,

$$\phi_1 : [0, 1] \times S^1 \to A_1,$$

is defined to be the union of the above parametrizations

$$\psi_1 \cup \psi_2 \cup \cdots \cup \psi_{2l-1}.$$

For the other annuli A_α, $\alpha = 2, \ldots, m$, define ϕ_α to be the

$$f^{\alpha-1}\phi_1 : [0, 1] \times S^1 \to A_\alpha.$$

(We are assuming $A_\alpha = f^{\alpha-1}(A_1)$.)

It is easy to check conditions (i)~(v) for these parametrizations

$$\phi_\alpha, \quad \alpha = 1, \ldots, m.$$

□

Definition 4.10. The parametrizations

$$\phi_\alpha : [0, 1] \times S^1 \to A_\alpha, \quad \alpha = 1, \ldots, m,$$

are called *superstandard parametrizations for* f. The homeomorphism

$$f : \bigcup_{\alpha=1}^{m} A_\alpha \to \bigcup_{\alpha=1}^{m} A_\alpha$$

is in superstandard form with respect to these parametrizations.

Corollary 4.1 (to Lemma 4.2). *(i) The equations describing*

$$f^m : A_\alpha \to A_\alpha$$

with respect to the superstandard parametrizations

$$\phi_\alpha : [0, 1] \times S^1 \to A_\alpha, \quad \alpha = 1, \ldots, m,$$

are uniquely determined by the series of numbers

$$(n_0, n_1, \ldots, n_l).$$

(ii) In particular, the valencies of $\partial_0 A_\alpha$ and $\partial_1 A_\alpha$ with respect to f are equal to (m, n_0, σ_0) and (m, n_l, σ_1), respectively, where σ_0 and σ_1 are determined by

$$0 \le \sigma_0 < n_0, \quad \sigma_0 \equiv n_1 \pmod{n_0}, \quad 0 \le \sigma_1 < n_l, \quad \sigma_1 \equiv n_{l-1} \pmod{n_l}.$$

The screw number of f in A_α is equal to

$$-\sum_{i=0}^{l-1} \frac{1}{n_i n_{i+1}}.$$

(iii) Conversely, the length of the $ARCH_\nu$ and the multiplicities

$$m_0, m_1, \ldots, m_l$$

of

$$D_0, S_1, \ldots, S_{l-1}, D_l$$

are determined by the valencies $(\mu_0, \lambda_0, \sigma_0)$ and $(\mu_1, \lambda_1, \sigma_1)$ of $\partial_0 A_\alpha$ and $\partial_1 A_\alpha$, together with the screw number of f in A_α, α being any of $1, \ldots, m$.

Proof. (i) The numbers δ_0/λ_0 and δ_1/λ_1 which describe the action of f^m on ∂A_α are determined by (λ_0, σ_0) and (λ_1, σ_1) but these two pairs are determined by

$$(n_0, n_1, \ldots, n_{l-1}, n_l)$$

as

$$\lambda_0 = n_0, \quad \lambda_1 = n_l,$$
$$0 \le \sigma_0 < n_0, \quad \sigma_0 \equiv n_1 \pmod{n_0},$$
$$0 \le \sigma_1 < n_l, \quad \sigma_1 \equiv n_{l-1} \pmod{n_l}.$$

See Lemma 3.1. The numbers d_i, $i = 1, \ldots, l-1$, which describe the action of f^m on the white annuli Y_{2i} $(i = 1, \ldots, l-1)$ might look uncontrolable. But one can find them by accumulating the twists $-1/(n_{i-1}n_i)$ starting from level $t = 0$:

$$-\frac{\delta_0}{\lambda_0} + \sum_{i=1}^{k} \frac{1}{n_{i-1}\, n_i} = \frac{d_k}{n_k}, \quad k = 1, \ldots, l-1.$$

(Remember our sign convention for twists. Cf. Corollary 2.2. See also Fig. 3.3.)

(ii) follows immediately from (i).

(iii) The $\gcd(m_{i-1}, m_i) = m$ is determined by the valency of $\partial_0 A_\alpha$ (or $\partial_1 A_\alpha$); $m = \mu_0 = \mu_1$. The numbers

$$(n_0,\ n_1,\ \ldots,\ n_l)$$

satisfy

$$n_0 = \lambda_0,\ \ n_l = \lambda_1,\ \ \ n_1 \equiv \sigma_0 \pmod{\lambda_0},\ \ \ n_{l-1} \equiv \sigma_1 \pmod{\lambda_1}$$

(by (ii)),

$$n_{i-1} + n_{i+1} \equiv 0 \pmod{n_i},\quad i = 1,\ \ldots,\ l-1$$

(by Proposition 3.1),

$$(n_{i-1} + n_{i+1})/n_i \geq 2,\quad i = 1,\ \ldots,\ l-1$$

(by the minimality assumption on S), and

$$\sum_{i=0}^{l-1} \frac{1}{n_i\, n_{i+1}} = |s(A_\alpha)|$$

by (ii).

Therefore, the sequence $(n_0, n_1, \ldots, n_l)$ is uniquely determined by (λ_0, σ_0), (λ_1, σ_1) and $s(A_\alpha)$. See Theorem 5.1 This proves (iii). □

4.3 Structure of π^{-1}(tail)

Let TL_ν, be a tail (of length $l \geq 1$). TL_ν contains an attaching bank D_0 and l spheres $S_1, \ldots, S_l$, which are trivial components of type I. Suppose

$$\{q_1\} = D_0 \cap S_1,\ \ \{q_i\} = S_{i-1} \cap S_i,\ \ \ i = 2,\ \ldots,\ l.$$

Let $\overline{N}_i$ be the closed nodal neighborhood of q_i $(i = 1, \dots, l)$, P_i the closed part in S_i $(i = 1, \dots, l)$. The multiplicities of

$$D_0,\ S_1,\ \dots,\ S_l$$

are denoted as usual:

$$m_0,\ m_1,\ \dots,\ m_l,$$

$m = \gcd(m_{i-1}, m_i)$ (being independent of i), and $n_i = m_i/m$.

The following lemma is proved by the same argument as Lemma 4.2. Here we use the polar coordinates (r, θ), $0 \le r \le 1, 0 \le \theta \le 2\pi$, for the unit disk D and set

$$\Delta' = \left\{(r, \theta) \mid 0 \le r \le \frac{1}{2l}\right\},$$

$$I_j \times S^1 = \left\{(r, \theta) \mid \frac{2l - j}{2l} \le r \le \frac{2l - j + 1}{2l}\right\}, \quad j = 1,\ 2,\ \dots,\ 2l - 1.$$

(N.B. $I_j \times S^1$ is on the outside of $I_{j+1} \times S^1$).

Lemma 4.3. *$\pi^{-1}(\mathrm{TL}_\nu)$ consists of m disks $\Delta_1, \Delta_2, \dots, \Delta_m$, and*

$$f : \pi^{-1}(\mathrm{TL}_\nu) \to \pi^{-1}(\mathrm{TL}_\nu)$$

cyclically permutes them:

$$f(\Delta_\alpha) = \Delta_{\alpha+1}, \quad \alpha = 1,\ \dots,\ m - 1, \quad f(\Delta_m) = \Delta_1.$$

There are parametrizations

$$\phi_\alpha : D \to \Delta_\alpha, \quad \alpha = 1,\ \dots,\ m,$$

such that

(i) $\pi^{-1}(\overline{N}_i) = \bigcup_{\alpha=1}^{m} \phi_\alpha(I_{2i-1} \times S^1), \quad i = 1,\ \dots,\ l.$

(ii) $\pi^{-1}(P_i) = \bigcup_{\alpha=1}^{m} \phi_\alpha(I_{2i} \times S^1), \quad i = 1,\ \dots,\ l - 1,$

(iii) $\pi^{-1}(P_l) = \bigcup_{\alpha=1}^{m} \phi_\alpha(\Delta'),$

(iv) $\phi_\alpha^{-1} f^m \phi_\alpha : I_{2i-1} \times S^1 \to I_{2i-1} \times S^1$ *is a linear twist with screw number*

$$-\frac{1}{n_{i-1} n_i}, \quad \alpha = 1,\ \dots,\ m, \quad i = 1,\ \dots,\ l.$$

(v) $\phi_\alpha^{-1} f^m \phi_\alpha : I_{2i} \times S^1 \to I_{2i} \times S^1$ *is a rotation sending*

$$(r, \theta) \quad \text{to} \quad \left(r, \theta + \frac{2\pi\delta_i}{n_i}\right),$$

where δ_i *is an integer coprime with* n_i, $\alpha = 1, \ldots, m$, $i = 1, \ldots, l-1$.

(vi) $\phi_\alpha^{-1} f^m \phi_\alpha : \Delta' \to \Delta'$ *is the identity.*

(vii) *Let* (m, λ, σ) *be the valency of* $\partial\Delta_\alpha$, *and let* δ *be determined by*

$$0 \leq \delta < \lambda, \quad \sigma\delta \equiv 1 \pmod{\lambda}.$$

Then

$$f^m \phi_\alpha(1, \theta) = \phi_\alpha\left(1, \theta + \frac{2\pi\delta}{\lambda}\right).$$

Definition 4.11. The parametrizations

$$\phi_\alpha : D \to \Delta_\alpha, \quad \alpha = 1, \ldots, m,$$

are called *superstandard parametrizations* for f. The homeomorphism

$$f : \bigcup_{\alpha=1}^{m} \Delta_\alpha \to \bigcup_{\alpha=1}^{m} \Delta_\alpha$$

is in superstandard form with respect to these parametrizations.

Corollary 4.2 (to Lemma 4.3). *(i) The equations describing*

$$f^m : \Delta_\alpha \to \Delta_\alpha$$

with respect to the superstandard parametrizations

$$\phi_\alpha : D \to \Delta_\alpha, \quad \alpha = 1, \ldots, m,$$

are uniquely determined by the series of numbers

$$(n_0, n_1, \ldots, n_l).$$

(ii) In particular, the valency of $\partial\Delta_\alpha$ *with respect to* f *is equal to*

$$(m, n_0, n_0 - n_1).$$

(iii) Conversely, the length of the TL_ν, *and the multiplicities*

$$m_0, m_1, \ldots, m_l$$

of

$$D_0,\ S_1,\ \ldots,\ S_l$$

are determined by the valency

$$(\mu,\ \lambda,\ \sigma)$$

of $\partial\Delta_\alpha$*,* α *being any of* $1, \ldots, m$*.*

Proof (of Corollary 4.2). First we prove (ii). This is nothing but an application of Lemma 3.1, but we must be careful about the orientation: In the present case, the valency of $\partial\Delta_\alpha$ is the valency of $\partial\overrightarrow{\Delta}_\alpha$(Cf. Chap. 1) unlike Lemma 3.1 where the boundary curves are oriented from the outside (see the explanation immediately before Corollary 2.1). This discrepancy results in the appearance of $n_0 - n_1$. This proves (ii).

To prove (i), we have only to note that the integers δ_k, $k = 1, \ldots, l-1$, which describe the action of f^m on $\phi_\alpha(I_{2k} \times S^1)$, are determined by the accumulation of the twists:

$$\sum_{i=k}^{l-1} \frac{1}{n_i\, n_{i+1}} = \frac{n_k - \delta_k}{n_k}, \quad k = 1,\ \ldots,\ l-1.$$

To prove (iii), we need a claim.

Claim (H). $n_0 > n_1 > \cdots > n_l = 1, \quad l \geq 1.$

Proof (of Claim). By Proposition 3.1, m_l divides m_{l-1}, and so

$$m_l = \gcd(m_{l-1}, m_l) = m,$$

implying $n_l = 1$. By the minimality assumption, we have

$$m_{l-1}/m_l \geq 2.$$

Thus $m_{l-1} > m_l$. Also by the minimality assumption

$$m_{i-1} + m_{i+1} \geq 2m_i, \quad i = 1,\ \ldots,\ l-1,$$

we have

$$m_{i-1} - m_i \ \geq\ m_i - m_{i+1}.$$

Starting with $m_{l-1} - m_l > 0$, we have inductively $m_{i-1} - m_i > 0, \quad i = 1,\ \ldots,\ l.$ □

Now assertion (iii) can be proved as follows:

$$m = \mu,\ \ n_0 = \lambda,\ \ n_1 = \lambda - \sigma,$$

(by (ii)), and

$$n_{i-1} + n_{i+1} \equiv 0 \pmod{n_i}, \quad i = 1, \dots, l-1,$$

(by Proposition 3.1). Then the sequence $(n_0, n_1, \dots, n_l)$ is uniquely determined by (λ, σ) using the Euclidean algorithm. □

4.4 Structure of π^{-1}(body)

Let BDY_ν be a body of S.

Lemma 4.4. *The Euler characteristic $\chi(\pi^{-1}(BDY_\nu))$ is negative, except for the body $BDY(m)$ depicted by Fig. 4.2. For this body the Euler characteristic of $\pi^{-1}(BDY(m))$ is zero.*

Proof. As was remarked after Lemma 4.1, BDY_ν is a union of its core part P_0 and the tails

$$\mathrm{TL}_1, \ \mathrm{TL}_2, \ \dots, \ \mathrm{TL}_k$$

attached to P_0. Let m_0 be the multiplicity of P_0. Then all the attaching disks

$$D_{(1)}, \ D_{(2)}, \ \dots, \ D_{(k)}$$

of

$$\mathrm{TL}_1, \ \mathrm{TL}_2, \ \dots, \ \mathrm{TL}_k$$

have the same multiplicity m_0. Let m_i be the multiplicity of the sphere $S_{(i)}$ in TL_i next to $D_{(i)}$ (i.e. such that $S_{(i)} \cap D_{(i)} \neq \emptyset$), $i = 1, \dots, k$. We denote the valency of a curve in $\pi^{-1}(\partial D_{(i)})$ by $(\mu_i, \lambda_i, \sigma_i)$. Then, by Corollary 4.2. (ii),

$$\mu_i = \gcd(m_0, m_i), \quad \lambda_i = m_0/\mu_i > 1.$$

(NB. $\lambda_i > 1$ because of Claim (H) in the proof of Corollary 4.2).

By Lemma 4.3, $\pi^{-1}(\mathrm{TL}_i)$ consists of μ_i disks.

Therefore,

$$\begin{aligned}
\chi(\pi^{-1}(BDY_\nu)) &= m_0\chi(P_0) + \sum_{i=1}^{k} \mu_i \\
&= m_0 \left(\chi(P_0) + \sum_{i=1}^{k} \frac{1}{\lambda_i} \right) \\
&= m_0 \left[\chi(\overline{P}_0) - \sum_{i=1}^{k} \left(1 - \frac{1}{\lambda_i} \right) \right],
\end{aligned}$$

Fig. 4.2 *BDY*(*m*)

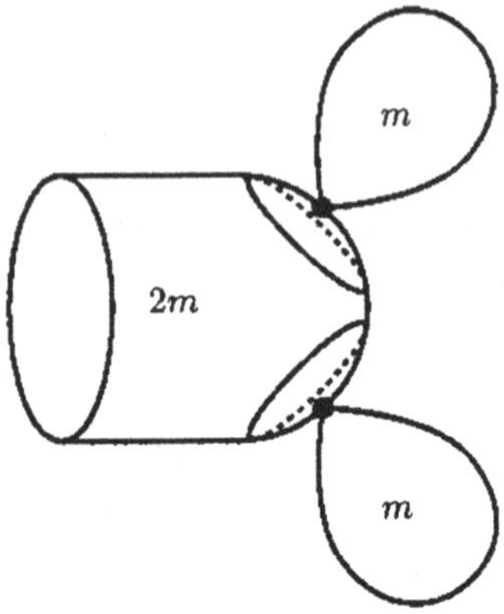

where $\overline{P}_0$ denotes the surface obtained from BDY_ν by replacing each tail TL_i $(i = 1, \ldots, k)$ by a disk.

Now assuming

$$\chi(\pi^{-1}(BDY_\nu)) \geq 0,$$

let us see what happens.

Since $1 - 1/\lambda_i > 0$, $\chi(\overline{P}_0)$ must be ≥ 0.
If $k = 0$, then

$$\chi(P_0) = \chi(\overline{P}_0) \geq 0.$$

This is impossible because P_0 is a core part. So $k > 0$, and $\chi(\overline{P}_0) > 0$.
If $\overline{P}_0$ is a sphere, $\pi^{-1}(BDY_\nu)$ must coincide with Σ_g. This is again impossible because $\chi(\Sigma_g) < 0$. Thus $\overline{P}_0$ is a disk. Since $1 - 1/\lambda_i \geq 1/2$, k is 1 or 2.
If $k = 1$, then P_0 is an annulus, contradicting $\chi(P_0) < 0$. Thus $k = 2$, and $\lambda_1 = \lambda_2 = 2$.

From

$$\lambda_i = m_0 / \gcd(m_0, m_i),$$

m_0 must be an even number, say $2m$, and

$$\gcd(m_0, m_i) = m.$$

Then $m_i = m$, and the length of TL_i must be $1, i = 1, 2$. (See Claim (H) in the proof of Corollary 4.2.) Thus we are led to the $BDY(m)$ of Fig. 4.2, for which

$$\chi(\pi^{-1}(BDY(m))) = 0$$

by the above formula. □

Corollary 4.3 (to Lemma 4.4). *Let $BDY(m)$ be the body shown in Fig. 4.2. Then $\pi^{-1}(BDY(m))$ consists of m annuli, $B'_1, B'_2, \ldots, B'_m$* (this notation for the annuli is to be consistent with the case of $Ch(\mathscr{A}_{\mathrm{sp}})$ of Chap. 3).

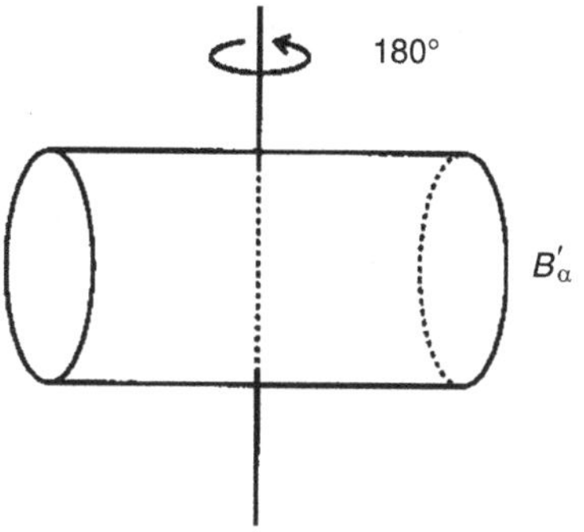

Fig. 4.3 $(f')^m : B'_\alpha \to B'_\alpha$ is a "180°-rotation"

$$f : \pi^{-1}(BDY(m)) \to \pi^{-1}(\mathrm{BDY}(m))$$

permutes these annuli cyclically, and is rel.∂ isotopic to f' such that, for each $\alpha = 1, \ldots, m$,

$$(f')^m : B'_\alpha \to B'_\alpha$$

is a 180° *rotation of B'_α around the axis in* Fig. 4.3.

The proof is left to the reader. (Consider the monodromy exponent in the proof of Proposition 3.1.)

Lemma 4.5. *Let*

$$ARCH_0 = D_0 \vee S_1 \vee \cdots \vee S_{l-1} \vee D_l$$

be an arch attached (by the bank D_l) to the body

$$BDY(m) = P_0 \cup \mathrm{TL}_1 \cup \mathrm{TL}_2$$

of Fig. 4.2. *Let*

$$m_0,\ m_1,\ \ldots,\ m_{l-1},\ m_l$$

be the multiplicities of

$$D_0,\ S_1,\ \ldots,\ S_{l-1},\ D_l.$$

Then

(i) $m_0 \geq m_1 \geq \ldots \geq m_l = 2m$, and the $\gcd(m_{i-1}, m_i)$ *is equal to $2m$ (independently of $i = 1, 2, \ldots, l$)*
(ii) $\pi^{-1}(ARCH_0 \cup BDY(m))$ consists of m annuli

$$A_1,\ A_2,\ \ldots,\ A_m,$$

and

$$f : \pi^{-1}(ARCH_0 \cup BDY(m)) \to \pi^{-1}(ARCH_0 \cup BDY(m))$$

cyclically permutes them. The m-th power

$$f^m : A_\alpha \to A_\alpha$$

interchanges the two boundary curves of the annulus, $\alpha = 1, \dots, m$.

(iii) *The screw number* $s(A_\alpha)$ *of* f *in* A_α *is equal to*

$$-2\sum_{i=0}^{l-1} \frac{1}{n_i\, n_{i+1}},$$

and the valency of a boundary curve of A_α *with respect to* f *is equal to*

$$(2m,\ n_0,\ \sigma_0),$$

where $n_i = m_i/2m$*, and* σ_0 *is determined by*

$$0 \le \sigma_0 < n_0, \quad \sigma_0 \equiv n_1 \pmod{n_0}.$$

Proof. (i) The multiplicity m_l of the bank D_l must be the same as the multiplicity of the core part P_0 of $BDY(m)$ to which D_l is attached. Thus $m_l = 2m$. We apply Proposition 3.1 to the irreducible component Θ_0 containing P_0. Then

$$m_{l-1} + m + m \equiv 0 \pmod{2m}.$$

This means

$$m_{l-1} \equiv 0 \pmod{2m}.$$

Thus

$$\gcd(m_{l-1}, m_l) = 2m, \quad \text{and} \quad m_{l-1} \ge 2m = m_l.$$

By the minimality assumption on S, we have

$$m_{i-1} - m_i \ge m_i - m_{i+1}, \quad i = 1, \dots, l-1.$$

Hence (i) follows.

(ii) By Lemma 4.2 and (i), $\pi^{-1}(ARCH_0)$ consists of $2m$ annuli

$$A'_1,\ A'_2,\ \dots,\ A'_{2m}$$

permuted by f cyclically. By Corollary 4.3, $\pi^{-1}(BDY(m))$ consists of m annuli

$$B'_1,\ B'_2,\ \dots,\ B'_m.$$

We may assume, for $\alpha = 1, \dots, m$,

$$A'_\alpha \cap B'_\alpha \ne \emptyset, \quad \text{and} \quad A'_{\alpha+m} \cap B'_\alpha \ne \emptyset.$$

Let A_α be the annulus

$$A'_\alpha \cup B'_\alpha \cup A'_{\alpha+m}, \quad \alpha = 1, 2, \dots, m.$$

Then $\pi^{-1}(ARCH_0 \cup BDY(m))$ is a disjoint union of m annuli,

$$A_1, A_2, \dots, A_m,$$

again permuted cyclically by f. Since f^m interchanges A'_α and $A'_{\alpha+m}$, it interchanges the boundary components of A_α.

(iii) The screw number of f in A_α is the sum of those in A'_α and $A'_{\alpha+m}$. These two are equal to

$$-\sum_{i=0}^{l-1} \frac{1}{n_i\, n_{i+1}}, \qquad \text{(Corollary 4.1.(ii))}.$$

Thus

$$s(A_\alpha) = s(A'_\alpha) + s(A''_\alpha) = -2\sum_{i=0}^{l-1} \frac{1}{n_i\, n_{i+1}}.$$

The assertion on the valency follows from Corollary 4.1(ii). □

With the same notation as in Lemma 4.5, we have the following

Lemma 4.6. *There are parametrizations*

$$\phi_\alpha : [0,1] \times S^1 \to A_\alpha, \quad (S^1 = \mathbf{R}/\mathbf{Z}), \quad \alpha = 1, \dots, m,$$

such that

(i) $\phi_\alpha|[0,1/3] \times S^1 : [0,1/3] \times S^1 \to A'_\alpha$ *and* $\phi_\alpha|[2/3,1] \times S^1 : [2/3,1] \times S^1 \to A'_{\alpha+m}$ *are superstandard parametrizations for*

$$f|\bigcup_{\alpha=1}^{2m} A'_\alpha,$$

in the sense of Lemma 4.2.

(ii) Let D_1^α, D_2^α *be the disks* ($\subset B'_\alpha$) *which are over the small tails* $\mathrm{TL}_1, \mathrm{TL}_2$ *of* $BDY(m)$*. Then* ϕ_α*, restricted to* $\phi_\alpha^{-1}(D_1^\alpha), \phi_\alpha^{-1}(D_2^\alpha)$*, provides superstandard parametrizations in the sense of* Lemma 4.3 *for* $f|\bigcup_{\alpha=1}^m D_1^\alpha$*, and* $f|\bigcup_{\alpha=1}^m D_2^\alpha$*, respectively.*

(iii) On the boundary ∂A_α*, the following holds*

$$f^m\, \phi_\alpha(0, x) = \phi_\alpha\left(1, -x - \frac{1}{4}s(A_\alpha)\right),$$

$$f^m\, \phi_\alpha(1, x) = \phi_\alpha\left(0, -x + \frac{1}{4}s(A_\alpha)\right).$$

The parametrizations ϕ_α are constructed separately for the three parts:

- the annular part
$$\bigcup_{\alpha=1}^{2m} A'_\alpha,$$
- the periodic part
$$\bigcup_{\alpha=1}^{m} (B'_\alpha - Int(D_1^\alpha) \cup Int(D_2^\alpha)),$$
  and
- the disk parts
$$\bigcup_{\alpha=1}^{m} D_1^\alpha, \ \bigcup_{\alpha=1}^{m} D_2^\alpha.$$

In the construction for $\bigcup_{\alpha=1}^{2m} A'_\alpha$, we can choose any "linear" coordinates for black annuli (see the proof of Lemma 4.2). Thus we can choose them so that the boundary behaviour of f^m satisfies (iii). (See Sublemma 1, and Corollary 2.4.) The details are left to the reader.

Definition 4.12. The parametrizations $\phi_\alpha : [0,1] \times S^1 \to A_\alpha, \alpha = 1, \ldots, m$, are called *superstandard parametrizations* for f. The homeomorphism
$$f : \bigcup_{\alpha=1}^{m} A_\alpha \to \bigcup_{\alpha=1}^{m} A_\alpha$$
is in superstandard form with respect to these parametrizations.

Remark 4.2. The equation of a special twist given in Chap. 2 was symmetric under the interchange of the parametrization
$$(t, x) \to (1-t, -x)$$
But the equations describing f in superstandard form (in the sense of Lemma 4.6) have no such symmetry in general; the symmetry remains only on the boundary action of f (Lemma 4.6(iii)). This asymmetry will cause no problem because all treatments on "superstandard level" will be done separately for the annular, periodic, and disk parts.

Corollary 4.4 (to Lemma 4.6). *(i) The equations describing $f^{2m}| \bigcup_{\alpha=1}^{2m} A'_\alpha$ with respect to the superstandard parametrizations*
$$\phi_\alpha \Big| \left[0, \frac{1}{3}\right] \times S^1 \cup \left[\frac{2}{3}, 1\right] \times S^1, \quad \alpha = 1, \ldots, m,$$

are uniquely determined by the series of numbers

$$(n_0, n_1, \ldots, n_l).$$

(ii) *Conversely, the length of $ARCH_0$ and the multiplicities*

$$m_0, m_1, \ldots, m_l,$$

of

$$D_0, S_1, \ldots, D_l$$

are determined by the valency (μ, λ, σ) of $\partial_0 A_\alpha$ and the screw number of f in A_α, α being any of $1, \ldots, m$.

This corollary is an interpretation of Corollary 4.1.

Definition 4.13. The body $BDY(m)$ depicted by Fig. 4.2 is called a *special body* (with multiplicity $2m$). The other bodies are called *ordinary bodies.*

Definition 4.14. The union of the special body $BDY(m)$ and an arch $ARCH_0$ attached to $BDY(m)$ is called a *special tail.* A special tail is denoted by SPL_ν (ν being a suffix).

Definition 4.15. An arch which is not attached to special bodies is called an *ordinary arch.*

4.5 Completion of the Proof of Theorem 4.1. (Uniqueness)

Remember that we had two pseudo-periodic maps

$$f, f' : \Sigma_g \to \Sigma_g$$

(of negative twist) in superstandard form, and that we were given their minimal quotients

$$\pi : \Sigma_g \to S \quad \text{and} \quad \pi' : \Sigma_g \to S'.$$

Assuming that f and f' are homotopic, we want to show that these quotients are essentially the same (in the sense of (i), (ii), (iii) of Theorem 4.1).

Let us assume that

$$\pi' : \Sigma_g \to S'$$

is constructed by the method of Chap. 3, as the union of

$$Ch(\mathscr{B}'),\ Ch(\mathscr{D}),\ Ch(\mathscr{A}_{\mathrm{ln}}) \text{ and } Ch(\mathscr{A}_{\mathrm{sp}}).$$

Then its structure is well understood. We examine the structure of

$$\pi : \Sigma_g \to S.$$

Let $ARCH^\circ$ (resp. BDY°, SPL) denote the collection of ordinary arches (resp. ordinary bodies, special tails) in S. S is decomposed as follows:

$$S = ARCH^\circ \cup BDY^\circ \cup \mathrm{SPL}.$$

By Corollary 4.1, $\pi^{-1}(ARCH^\circ)$ is a disjoint union of non-amphidrome annuli, in which the screw number of f is non-zero. By Lemma 4.4, $\pi^{-1}(BDY^\circ)$ is a disjoint union of compact surfaces, which have negative Euler characteristics. Finally by Lemma 4.5, $\pi^{-1}(\mathrm{SPL})$ is a disjoint union of amphidrome annuli, in which the screw number of f is non-zero.

Therefore, the decomposition of Σ_g

$$\Sigma_g = \pi^{-1}(ARCH^\circ) \cup \pi^{-1}(BDY^\circ) \cup \pi^{-1}(\mathrm{SPL})$$

corresponds to the decomposition by the precise system of cut curves subordinate to

$$f : \Sigma_g \to \Sigma_g$$

(Chap. 1).

Let us deform f and f' (which are in superstandard form) back into standard forms $\overline{f}$ and $\overline{f}'$.

N.B. Linear and special twists for $\overline{f}$ will be considered with respect to the superstandard parametrizations provided by Lemmmas 4.2 and 4.6. In the annuli where $\overline{f}$ are to be special twists, the property (iii) of the parametrizations (symmetry on the boundary) imposed by Lemma 4.6 assures that f can be deformed into special twists in the annuli without changing the parametrizations. See the proof of Theorem 2.4(i). "Linear" and "special" parametrizations for $\overline{f}'$ already exist in A_{ln} and A_{sp}, because f' was constructed starting from $\overline{f}'$ in standard form. See Chap. 3.

The standard forms $\overline{f}$ and $\overline{f}'$ preserve the decompositions

$$\Sigma_g = \pi^{-1}(ARCH^\circ) \cup \pi^{-1}(BDY^\circ) \cup \pi^{-1}(\mathrm{SPL}),$$

and

$$\Sigma_g = \mathscr{A}_{\mathrm{ln}} \cup \mathscr{B} \cup \mathscr{A}_{\mathrm{sp}},$$

where $\mathscr{B} = \Sigma_g - Int(\mathscr{A}_{\mathrm{ln}}) \cup Int(\mathscr{A}_{\mathrm{sp}})$, respectively. See the proof of Theorem 2.1(i), at the end of Chap. 2.

By our assumption of Theorem 4.1, $\overline{f}$ and $\overline{f}'$ are mutually homotopic. Then by Theorem 2.1 (ii) (Uniqueness of standard form), there is a homeomorphism

$$h : \Sigma_g \to \Sigma_g$$

isotopic to the identity, such that

$$\overline{f} = h^{-1}\overline{f}'h.$$

The homeomorphism h preserves the decomposition i.e.

$$h(\pi^{-1}(ARCH^\circ)) = \mathscr{A}_{\text{ln}}, \quad h(\pi^{-1}(BDY^\circ)) = \mathscr{B}, \quad h(\pi^{-1}(\text{SPL})) = \mathscr{A}_{\text{sp}}$$

(See the proof of Theorem 2.1 (ii).)

Let us look at the linear part $h^{-1}(\pi^{-1}(ARCH^\circ))$ more closely. Let $ARCH_\nu^\circ$ be an ordinary arch. By Lemma 4.2, $\pi^{-1}(ARCH_\nu^\circ)$ is a disjoint union of annuli

$$A_1, \ A_2, \ \ldots, \ A_m$$

which are permuted by f cyclically:

$$f(A_\alpha) = A_{\alpha+1}, \quad \alpha = 1, \ldots, m-1, \quad f(A_m) = A_1.$$

Thus $\pi^{-1}(ARCH_\nu^\circ)$ is a cyclic orbit of the permutation on the annuli in $\pi^{-1}(ARCH^\circ)$ caused by f.

Denote $h(A_\alpha)$ by A'_α. The disjoint union

$$\mathscr{A}_{\text{ln}}^{(\nu)} = A'_1 \cup A'_2 \cup \cdots \cup A'_m$$

is a cyclic orbit of the permutation on the annuli in $\mathscr{A}_{\text{ln}}$ caused by f'.

Take an annulus A_1 in $\pi^{-1}(ARCH_\nu^\circ)$. By Corollary 4.1 (i), the equations describing

$$f^m : A_1 \to A_1$$

in superstandard form are uniquely determined by the series of numbers

$$(n_0, \ n_1, \ \ldots, \ n_l).$$

The same thing can be said on A'_1 and

$$(f')^m : A'_1 \to A'_1,$$

i.e. the equations describing

$$(f')^m : A'_1 \to A'_1$$

in superstandard form are uniquely determined by the series of numbers

$$(n'_0, \ n'_1, \ \ldots, \ n'_{l'})$$

carrying the same meaning as

$$(n_0, n_1, \ldots, n_l).$$

By Corollary 4.1 (iii), the numbers

$$(n_0, n_1, \ldots, n_l)$$

are determined by the valencies of $\partial_0 A_1$ and $\partial_1 A_1$, together with the screw number of f in A_1. The numbers

$$(n'_0, n'_1, \ldots, n'_{l'})$$

are determined in the same fashion (again by Corollary 4.1 (iii)). But we have

$$\overline{f} = h^{-1}\overline{f}'h,$$

which implies that the corresponding valencies and screw numbers are equal.

Therefore,

$$(n_0, n_1, \ldots, n_l) = (n'_0, n'_1, \ldots, n'_{l'}), \quad l = l',$$

and the equations for

$$f^m : A_1 \to A_1$$

and

$$(f')^m : A'_1 \to A'_1$$

coincide.

When we constructed (in Chap. 2) the homeomorphism

$$h : \Sigma_g \to \Sigma_g,$$

we adjusted so that

$$h|A_1 : A_1 \to A'_1$$

is linear with respect to the parametrizations

$$\phi_1 : [0, 1] \times S^1 \to A_1$$

and

$$\phi'_1 : [0, 1] \times S^1 \to A'_1$$

with which the linearity of $\overline{f}$ and $\overline{f}'$ are constructed, respectively. Also $h|A_1$ preserves the t-levels of these parametrizations. (See the proof of Lemma 2.2.)

Remember that these parametrizations ϕ_1, ϕ'_1 are superstandard parametrizations for f and f'. Since the equations for

$$f^m : A_1 \to A_1$$

and

$$(f')^m : A'_1 \to A'_1$$

(with respect to ϕ_1 and ϕ'_1) coincide, and

$$h|A_1 : A_1 \to A_1$$

is linear with respect to ϕ_1 and ϕ'_1, we have $f^m = h^{-1}(f')^m h$ on A_1.

Moreover, the two homeomorphisms

$$f, \quad h^{-1} f' h : (\pi^{-1}(ARCH^\circ_\nu), \partial\pi^{-1}(ARCH^\circ_\nu)) \to$$
$$(\pi^{-1}(ARCH^\circ_\nu), \partial\pi^{-1}(ARCH^\circ_\nu))$$

are isotopic *rel.* $\partial\pi^{-1}(ARCH^\circ_\nu)$, because f (resp. f') is isotopic to $\overline{f}$ (resp. $\overline{f}'$) *rel.* $\partial\pi^{-1}(ARCH^\circ_\nu)$ (resp. *rel.* $\partial\mathscr{A}^{(\nu)}_{\mathrm{ln}}$),
and

$$\overline{f} = h^{-1}\overline{f}' h.$$

Then using this isotopy and the fact that $f^m = h^{-1}(f')^m h$, we can proceed as in the second half of the proof of Theorem 2.3 (ii), and can find an isotopy

$$h'_\tau : \pi^{-1}(ARCH^\circ_\nu) \to \pi^{-1}(ARCH^\circ_\nu), \quad 0 \le \tau \le 1$$

such that

$$h'_0 = \mathrm{id}$$
$$h'_\tau|\partial\pi^{-1}(ARCH^\circ_\nu) = \mathrm{id}$$
$$f = (h'_1)^{-1} h^{-1} f' h\,(h'_1).$$

Replacing h with $h\,h'_1$, we may assume $f = h^{-1} f' h$ on $\pi^{-1}(ARCH^\circ_\nu)$.

Now it is easy to see that h projects to a numerical homeomorphism H_ν : $ARCH^\circ_\nu \to Ch(\mathscr{A}^{(\nu)}_{\mathrm{ln}})$ which makes the diagram commute:

$$\begin{array}{ccc} \pi^{-1}(ARCH^\circ_\nu) & \xrightarrow{\;h\;} & \mathscr{A}^{(\nu)}_{\mathrm{ln}} \\ \pi\downarrow & & \downarrow\pi' \\ ARCH^\circ_\nu & \xrightarrow{\;H_\nu\;} & Ch(\mathscr{A}^{(\nu)}_{\mathrm{ln}}) \end{array}$$

Taking the disjoint unions $\bigcup_\nu \pi^{-1}(ARCH^\circ_\nu)$, $\bigcup_\nu \mathscr{A}^{(\nu)}_{\mathrm{ln}}$, we have a homeomorphism

$$h : \pi^{-1}(ARCH^\circ) \to \mathscr{A}_{\mathrm{ln}}$$

and a numerical homeomorphism $H : ARCH^\circ \to Ch(\mathscr{A}_{\text{ln}})$ which make the diagram commute:

$$\begin{array}{ccc} \pi^{-1}(ARCH^\circ) & \xrightarrow{h} & \mathscr{A}_{\text{ln}} \\ \pi \downarrow & & \downarrow \pi' \\ ARCH^\circ & \xrightarrow{H} & Ch(\mathscr{A}_{\text{ln}}) \end{array}$$

We are done in the part $\pi^{-1}(ARCH^\circ)$.

To carry out the construction in the part $\pi^{-1}(BDY^\circ)$, divide $\pi^{-1}(BDY^\circ)$ into periodic parts and rotational parts, the latter being over the tails, adjust h so that it preserves this decomposition, and apply Lemma 4.3 and Corollary 4.2 to the rotational parts.

Finally in $\pi^{-1}(\text{SPL})$, the construction is done separately: in the linear part over the "special" arches, in the periodic parts over the closed parts, and in the rotational parts over the small tails. In these constructions, use the parametrizations provided by Lemma 4.6.

Putting together these results, we obtain a homeomorphism $h : \Sigma_g \to \Sigma_g$ and a numerical homeomorphism $H : S \to S'$ such that

(i) h is isotopic to the identity,
(ii) $f = h^{-1} f' h$, and
(iii) $\pi' h = H \pi$.

This completes the proof of Theorem 4.1. □

4.6 General Definition of Minimal Quotient

Let

$$f : \Sigma_g \to \Sigma_g$$

be a pseudo-periodic map of negative twist. By Theorem 3.1 (Existence), f is isotopic to f' in superstandard form which has its minimal quotient

$$\pi : \Sigma_g \to S.$$

By Theorem 4.1 (Uniqueness), this quotient

$$\pi : \Sigma_g \to S$$

is independent of the choice of the superstandard form f' of f.

Thus the following definition makes sense:

Definition 4.16. The above quotient

$$\pi : \Sigma_g \to S$$

is called *the minimal quotient* of

$$f : \Sigma_g \to \Sigma_g.$$

It is also called *the minimal quotient of the mapping class* $[f]$ ($\in M_g$) to which f belongs.

Notation
To make f explicit in the notation of its minimal quotient, we will denote it by

$$\pi : \Sigma_g \to S[f].$$

Proposition 4.1. *Let*

$$f : \Sigma_g \to \Sigma_g$$

be a pseudo-periodic map of negative twist, then the following diagram homotopically commutes:

$$\begin{array}{ccc} \Sigma_g & \xrightarrow{f} & \Sigma_g \\ {\scriptstyle \pi}\downarrow & & \downarrow{\scriptstyle \pi} \\ S[f] & \xrightarrow[id.\,(=)]{} & S[f] \end{array}$$

The proof is immediate.

4.7 Conjugacy Invariance

Theorem 4.2. *Let*

$$f_1 : \Sigma_g^{(1)} \to \Sigma_g^{(1)}$$

and

$$f_2 : \Sigma_g^{(2)} \to \Sigma_g^{(2)}$$

be pseudo-periodic maps of negative twist. Suppose there exists a homeomorphism

$$h : \Sigma_g^{(1)} \to \Sigma_g^{(2)}$$

such that $f_1 \simeq h^{-1} f_2 h$ *(homotopic). Then there exist a homeomorphism*

$$h' : \Sigma_g^{(1)} \to \Sigma_g^{(2)}$$

and a numerical homeomorphism

$$H : S[f_1] \to S[f_2]$$

such that

(i) h' *is isotopic to* h, *and*
(ii) $\pi_2 h' = H\pi_1$, *i.e. the following diagram commutes*:

$$\begin{array}{ccc} \Sigma_g^{(1)} & \xrightarrow{h'} & \Sigma_g^{(2)} \\ {\scriptstyle \pi_1}\downarrow & & \downarrow{\scriptstyle \pi_2} \\ S[f_1] & \xrightarrow{H} & S[f_2] \end{array}$$

Here

$$\pi_1 : \Sigma_g^{(1)} \to S[f_1]$$

and

$$\pi_2 : \Sigma_g^{(2)} \to S[f_2]$$

denote the respective minimal quotients.

(iii) Suppose that

$$f_i' : \Sigma_g^{(i)} \to \Sigma_g^{(i)}$$

is a pseudo-periodic map in superstandard form which is isotopic to

$$f_i : \Sigma_g^{(i)} \to \Sigma_g^{(i)}$$

and that f_i' *gives the minimal quotient*

$$\pi_i : \Sigma_g^{(i)} \to S[f_i], \quad i = 1,\ 2.$$

Then $f_1' = (h')^{-1} f_2' h'$.

Proof. Let f_1' and f_2' be the superstandard forms of f_1 and f_2, respectively. There is a numerical homeomorphism

$$H_2 : S[h^{-1} f_2' h] \to S[f_2']$$

such that the following diagram commutes:

$$\begin{array}{ccc} \Sigma_g^{(1)} & \xrightarrow{h} & \Sigma_g^{(2)} \\ {\scriptstyle \pi'}\downarrow & & \downarrow{\scriptstyle \pi_2} \\ S[h^{-1} f_2' h] & \xrightarrow[H_2]{} & S[f_2'] = S[f_2]\ . \end{array}$$

H_2 is simply the projection of h under the identifications $\pi'$and π_2.
Since $f_1' \simeq h^{-1} f_2' h$ (homotopic), there exist a homeomorphism

$$h_1 : \Sigma_g^{(1)} \to \Sigma_g^{(1)}$$

and a numerical homeomorphism

$$H_1 : S[f_1'] \to S[h^{-1} f_2' h]$$

such that h_1 is isotopic to the identity and the following diagram commutes (Theorem 4.1):

$$\begin{array}{ccc} \Sigma_g^{(1)} & \xrightarrow{h_1} & \Sigma_g^{(2)} \\ {\scriptstyle \pi_1}\downarrow & & \downarrow{\scriptstyle \pi'} \\ S[f_1'] & \xrightarrow[H_1]{} & S[h^{-1} f_2' h]\,. \end{array}$$

Also we have $f_1' = (h_1)^{-1} (h^{-1} f_2' h)\, h_1$.
Define a homeomorphism

$$h' : \Sigma_g^{(1)} \to \Sigma_g^{(2)}$$

to be hh_1 and a numerical homeomorphism $H : S[f_1] \to S[f_2]$ to be $H_2 H_1$, then h' is isotopic to h and the following diagram commutes:

$$\begin{array}{ccc} \Sigma_g^{(1)} & \xrightarrow{h'} & \Sigma_g^{(2)} \\ {\scriptstyle \pi_1}\downarrow & & \downarrow{\scriptstyle \pi_2} \\ S[f_1] & \xrightarrow[H]{} & S[f_2] \end{array}$$

□

Definition 4.17. The pinched coverings $\pi : \Sigma \to S$ and $\pi' : \Sigma' \to S'$ are *isomorphic* if there exist a homeomorphism

$$h : \Sigma \to \Sigma'$$

and a numerical homeomorphism

$$H : S \to S'$$

such that $\pi' h = H \pi$.

Corollary 4.5 (to Theorem 4.2). *The isomorphism class of the minimal quotient*

$$\pi : \Sigma_g \to S[f]$$

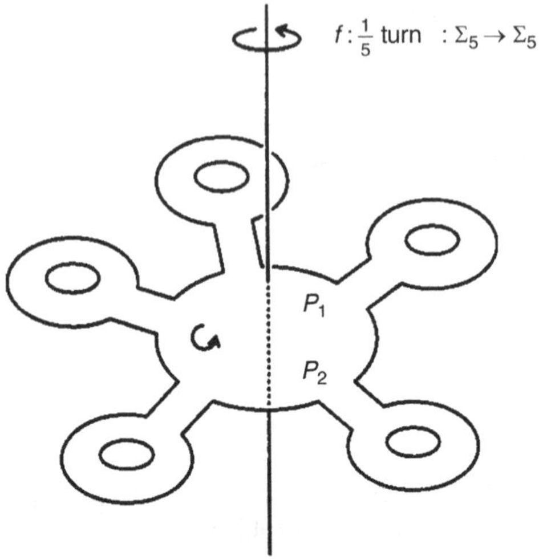

Fig. 4.4 $1/5$-turn $f : \Sigma_5 \to \Sigma_5$

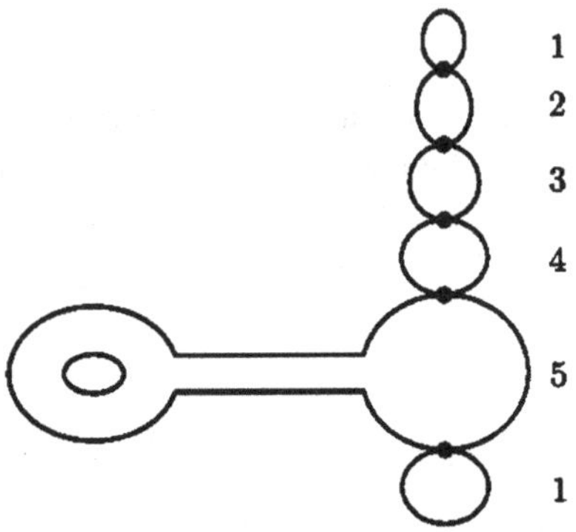

Fig. 4.5 The minimal quotient of the $1/5$-turn $f : \Sigma_5 \to \Sigma_5$

is a conjugacy invariant of $[f] \in \mathscr{M}_g$*, i.e. the isomorphism class is independent of the choice of the representative of the conjugacy class of* $[f]$ *in the group* $\mathscr{M}_g$*.*

Example 4.1. $\frac{1}{5}$*-turn of* Σ_5 $f : \Sigma_5 \to \Sigma_5$, (See Fig. 4.4) ,
valency of $P_1 = (1, 5, 1)$ (NB: $\sigma = 1, \delta = 1$):

$$n_0 = 5, \;\; n_1 = 5 - 1 = 4, \;\; n_2 = 3, \;\; n_3 = 2, \;\; n_4 = 1, \quad \text{Lemma 3.2},$$

valency of $P_2 = (1, 5, 4)$ (NB: $\sigma = 4, \delta = 4$):

$$n_0 = 5, \;\; n_1 = 5 - 4 = 1, \quad \text{Lemma 3.2}.$$

The minimal quotient $S[f]$: see Fig. 4.5.

Fig. 4.6 2/5-turn $f : \Sigma_5 \to \Sigma_5$

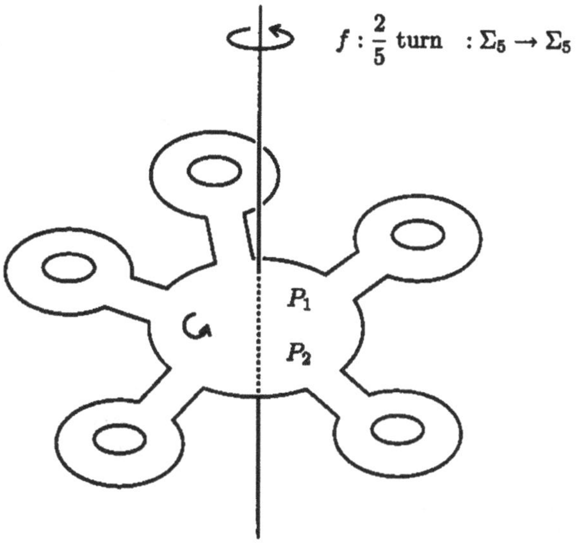

Fig. 4.7 The minimal quotient of the 2/5-turn $f : \Sigma_5 \to \Sigma_5$

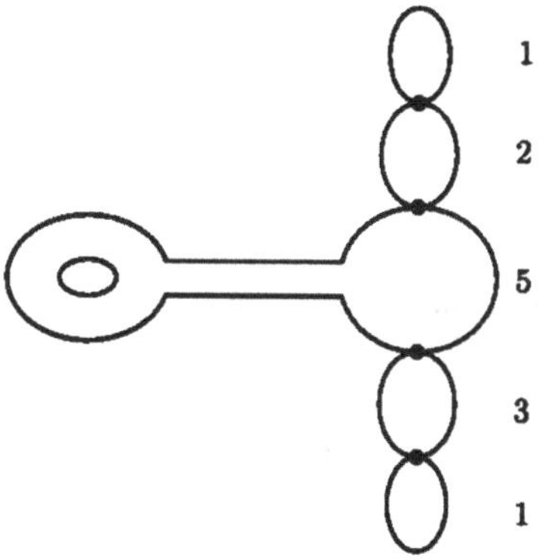

Example 4.2. $\frac{2}{5}$*-turn of* Σ_5 $\;f : \Sigma_5 \to \Sigma_5$, (See Fig. 4.6), valency of $P_1 = (1, 5, 3)$ (NB: $\sigma = 3, \delta = 2$):

$$n_0 = 5, \;\; n_1 = 5 - 3 = 2, \;\; n_2 = 1,$$

valency of $P_2 = (1, 5, 2)$, (NB: $\sigma = 2, \delta = 3$):

$$n_0 = 5, \;\; n_1 = 5 - 2 = 3, \;\; n_2 = 1.$$

The minimal quotient $S[f]$: see Fig. 4.7.

Example 4.3. (See Figs. 4.8 and 4.9). $f|B_1 \simeq$ id; $f|(B_2 - C_2) \simeq 180°$rotation; C_2 is amphidrome: $S(C_2) = -2$;

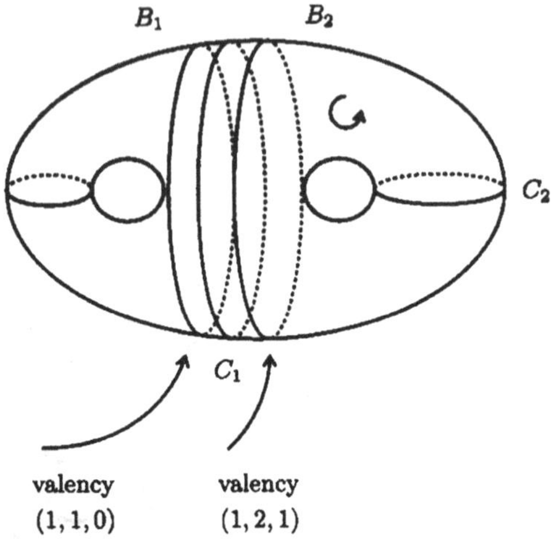

Fig. 4.8 Example 4.3

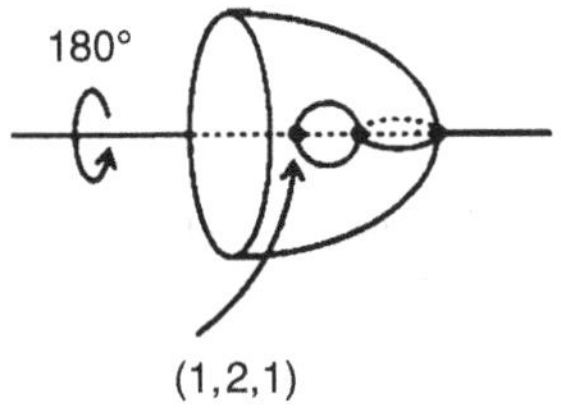

Fig. 4.9 $f|(B_2 - C_2)$ is a "180°"-rotation

C_1 is non-amphidrome: $S(C_1) = -3/2$ with $(n_0, n_1, n_2) = (1, 1, 2)$ (see Theorem 5.1). The minimal quotient $S[f]$ is depicted in Fig. 4.10.

Proposition 4.2. *Let*

$$\pi : \Sigma_g \to S$$

be a generalized quotient of a pseudo-periodic map

$$f : \Sigma_g \to \Sigma_g.$$

Let BDY_0 be a body in S. Let P_0 be the core part of BDY_0, Θ_0 the irreducible component (with the multiplicity m_0) containing P_0. Suppose that Θ_0 intersects the other irreducible components in k points, $p_1, p_2, \ldots, p_k$, and that the multiplicity of the irreducible component which intersects Θ_0 at p_i is m_i $(i = 1, 2, \ldots, k)$.

(i) *If P_0 is of genus* 0 (namely, a sphere with a number of disks deleted), *then the number of the connected components of $\pi^{-1}(BDY_0)$ is equal to* $\gcd(m_0, m_1, \ldots, m_k)$.
(ii) *If P_0 is of genus $\geq$ 1, then the number of the connected components of $\pi^{-1}(BDY_0)$ is a common divisor of $m_0, m_1, \ldots, m_k$.*
(iii) *In both cases, $f : \pi^{-1}(BDY_0) \to \pi^{-1}(BDY_0)$ permutes the connected components cyclically.*

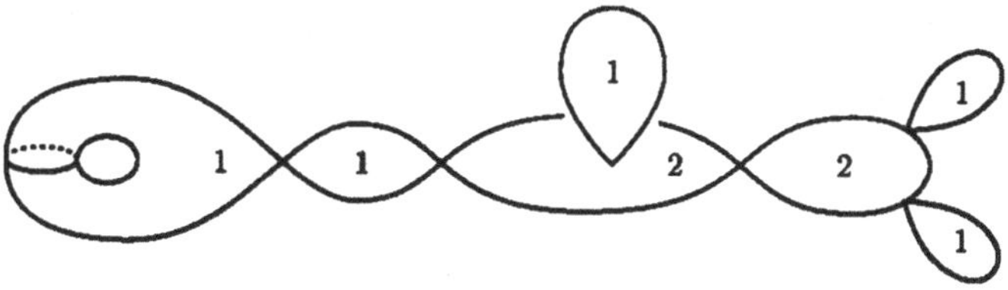

Fig. 4.10 The minimal quotient $S[f]$ of Example 4.3

The numbers of the connected components of $\pi^{-1}(BDY_0)$ and $\pi^{-1}(P_0)$ are equal, so we have only to consider $\pi^{-1}(P_0)$. The proof is a simple application of the monodromy exponent

$$\omega : H_1(P_0) \to \mathbf{Z}/m_0$$

considered in the proof of Proposition 3.1. See also [[51], Sect. 2].

If P_0 has positive genus, the behaviour of ω "around the genus" cannot be controlled by the multiplicities downstairs. The ambiguity thus arises in (ii). Assertion (iii) is an immediate consequence of the definition of a generalized quotient.

The details are left to the reader.

We conclude Chap. 4 with the following proposition, which shows that the number of the connected components of the preimage of a body is a strong local invariant.

Let

$$f_1 : \Sigma_g^{(1)} \to \Sigma_g^{(1)}$$

and

$$f_2 : \Sigma_g^{(2)} \to \Sigma_g^{(2)}$$

be pseudo-periodic maps of negative twist, each being in superstandard form. Let

$$\pi_1 : \Sigma_g^{(1)} \to S^{(1)}$$

and

$$\pi_2 : \Sigma_g^{(2)} \to S^{(2)}$$

be the respective minimal quotients.

We assume that there is a numerical homeomorphism

$$H : S^{(1)} \to S^{(2)}.$$

Up to isotopy, H may be assumed to preserve the systems of closed nodal neighborhoods. Let $BDY_0^{(1)}$ be a body in $S^{(1)}$, $BDY_0^{(2)} = H(BDY_0^{(1)})$ the image of $BDY_0^{(1)}$.

Proposition 4.3. *If the numbers of the connected components of* $\pi_1^{-1}(BDY_0^{(1)})$ *and* $\pi_2^{-1}(BDY_0^{(2)})$ *are equal, then there is a homeomorphism*

$$h' : \pi_1^{-1}(BDY_0^{(1)}) \to \pi_2^{-1}(BDY_0^{(2)})$$

and a numerical homeomorphism

$$H' : BDY_0^{(1)} \to BDY_0^{(2)}$$

such that

(i) the following diagram commutes:

$$\begin{array}{ccc} \pi_1^{-1}(BDY_0^{(1)}) & \xrightarrow{h'} & \pi_2^{-1}(BDY_0^{(2)}) \\ f_1 \downarrow & & \downarrow f_2 \\ \pi_1^{-1}(BDY_0^{(1)}) & \xrightarrow{h'} & \pi_2^{-1}(BDY_0^{(2)}) \\ \pi_1 \downarrow & & \downarrow \pi_2 \\ BDY_0^{(1)} & \xrightarrow{H'} & BDY_0^{(2)} \end{array}$$

and

(ii) $H'|_{\partial BDY_0^{(1)}} = H|_{\partial BDY_0^{(1)}}$.

Proof. If $BDY_0^{(1)}$ is a special body (Fig. 4.2), the proposition follows from Corollary 4.3. Thus we assume that $BDY_0^{(1)}$ is an ordinary body. Let $P_0^{(1)}$ (resp. $P_0^{(2)}$) be the core part of $BDY_0^{(1)}$ (resp. $BDY_0^{(2)}$).

$$H : S^{(1)} \to S^{(2)}$$

preserves the multiplicities, so it also preserves the valencies of the boundary curves of $P_0^{(1)}$ and $P_0^{(2)}$. (See Lemma 3.1. Also see the proof of Proposition 3.1 for the explanation of the valencies of the boundary curves of $P_0^{(i)}$.) Thus there are the same number of circles over a boundary curve Γ_i of $P_0^{(1)}$ and over the boundary curve $H(\Gamma_i)$ of $P_0^{(2)}$. Let

$$Q_1^{(1)},\ Q_2^{(1)},\ \ldots,\ Q_m^{(1)} \quad (\text{resp. } Q_1^{(2)},\ Q_2^{(2)},\ \ldots,\ Q_m^{(2)})$$

be the connected components of

$$\pi_1^{-1}(P_0^{(1)}) \quad (\text{resp. } \pi_2^{-1}(P_0^{(2)})),$$

which are cyclically permuted by f_1 (resp. f_2). By the above remark, $Q_\alpha^{(1)}$ and $Q_\beta^{(2)}$ have the same number of boundary components, $\alpha, \beta = 1, \dots, m$. Moreover, since $P_0^{(1)}$ and $P_0^{(2)}$ are homeomorphic and have the same multiplicity, say m_0, we have

$$\chi(\pi_1^{-1}(P_0^{(1)})) = \chi(\pi_2^{-1}(P_0^{(2)})) \quad (< 0),$$

implying

$$\chi(Q_\alpha^{(1)}) = \chi(Q_\beta^{(2)}), \quad \alpha, \beta = 1, \dots, m.$$

Therefore, $Q_\alpha^{(1)}$ and $Q_\beta^{(2)}$ are homeomorphic, $\alpha, \beta = 1, \dots, m$. We may assume

$$f_1(Q_\alpha^{(1)}) = Q_{\alpha+1}^{(1)}, \quad \alpha = 1, \dots, m-1,$$

$$f_1(Q_m^{(1)}) = Q_1^{(1)},$$

and similarly for $Q_\beta^{(2)}$.

Apply Theorem 1.3 to

$$f_1^m : Q_1^{(1)} \to Q_1^{(1)}$$

and

$$f_2^m : Q_1^{(2)} \to Q_1^{(2)},$$

both of which are periodic maps of order m_0/m. Then there exist a homeomorphism

$$H' : P_0^{(1)} \to P_0^{(2)}$$

and a homeomorphism

$$h_1 : Q_1^{(1)} \to Q_1^{(2)}$$

such that

1. the following diagram commutes:

$$\begin{array}{ccc}
Q_1^{(1)} & \xrightarrow{h_1} & Q_1^{(2)} \\
{\scriptstyle f_1^m}\downarrow & & \downarrow{\scriptstyle f_2^m} \\
Q_1^{(1)} & \xrightarrow{h_1} & Q_1^{(2)} \\
{\scriptstyle \pi_1}\downarrow & & \downarrow{\scriptstyle \pi_2} \\
P_0^{(1)} & \xrightarrow{h'} & P_0^{(2)}
\end{array}$$

and

2. $H'|_{\partial P_0^{(1)}} = H|_{\partial P_0^{(1)}}$.

For $\alpha = 2, \ldots, m$, define

$$h_\alpha : Q_\alpha^{(1)} \to Q_\alpha^{(2)}$$

by setting

$$h_\alpha = (f_2)^{\alpha-1} h_1 (f_1)^{1-\alpha},$$

and define

$$h : \pi_1^{-1}(P_0^{(1)}) \to \pi_2^{-1}(P_0^{(2)})$$

to be the union

$$h_1 \cup h_2 \cup \cdots \cup h_m.$$

Then it is easy to check on $\pi_1^{-1}(P_0^{(1)})$ that

$$hf_1 = f_2 h \quad \text{and} \quad H'\pi_1 = \pi_2 h.$$

Suppose $P_0^{(1)}$ has k tails

$$\mathrm{TL}_1^{(1)}, \ \mathrm{TL}_2^{(1)}, \ \ldots, \ \mathrm{TL}_k^{(1)},$$

$\mathrm{TL}_\nu^{(1)}$ being attached to $\partial_\nu P_0^{(1)}$. Set

$$\mathrm{TL}_\nu^{(2)} \ = \ H(\mathrm{TL}_\nu^{(1)}), \quad \nu = 1, \ \ldots, \ k.$$

Let

$$(\mu_\nu, \ \lambda_\nu, \ \sigma_\nu)$$

be the valencies of

$$\partial_\nu P_0^{(1)}, \quad \nu = 1, \ \ldots, \ k.$$

Since

$$H : S^{(1)} \to S^{(2)}$$

preserves these valencies,

$$\partial_\nu P_0^{(2)} \ \ (= \ H(\partial_\nu P_0^{(1)}))$$

has the same valency

$$(\mu_\nu, \ \lambda_\nu, \ \sigma_\nu), \quad \nu = 1, \ \ldots, \ k.$$

By Lemma 4.3 and Corollary 4.2,

$$\pi_1^{-1}(\mathrm{TL}_\nu^{(1)}) \ \text{(resp. } \pi_2^{-1}(\mathrm{TL}_\nu^{(2)}))$$

consists of μ_ν disks

$$\Delta^{(1)}_{\nu,\alpha} \quad (\text{resp. } \Delta^{(2)}_{\nu,\alpha}), \qquad \alpha = 1, \ldots, \mu_\nu,$$

permuted cyclically by f_1 (resp. f_2), and

$$(f_1)^{\mu_\nu} : \Delta^{(1)}_{\nu,\alpha} \to \Delta^{(1)}_{\nu,\alpha}$$

and

$$(f_2)^{\mu_\nu} : \Delta^{(2)}_{\nu,\alpha} \to \Delta^{(2)}_{\nu,\alpha}$$

have the identical superstandard expressions, $\alpha = 1, \ldots, \mu_\nu$. Thus, defining

$$h_{\nu,\alpha} : \Delta^{(1)}_{\nu,\alpha} \to \Delta^{(2)}_{\nu,\alpha}$$

to be the "identity" with respect to the superstandard parametrizations ($\alpha = 1, \ldots, \mu_\nu$), we can extend

$$h : \pi_1^{-1}(P_0^{(1)}) \to \pi_2^{-1}(P_0^{(2)})$$

to

$$h' = h \cup (\bigcup_{\nu,\alpha} h_{\nu,\alpha}) : \pi_1^{-1}(BDY_0^{(1)}) \to \pi_1^{-1}(BDY_0^{(2)}),$$

which projects to a homeomorphism (again denoted by)

$$H' : BDY_0^{(1)} \to BDY_0^{(2)}.$$

Obviously they satisfy

$$h' f_1 = f_2 h' \quad \text{and} \quad H'|_{\partial BDY_0^{(1)}} = H|_{\partial BDY_0^{(1)}}.$$

□

Chapter 5
A Theorem in Elementary Number Theory

In this chapter we will prove the following theorem which was essential to the arguments in Chaps. 3 and 4.

Theorem 5.1. *Let* $(\lambda_0, \sigma_0, \delta_0)$ *and* $(\lambda_1, \sigma_1, \delta_1)$ *be two triples of integers, each satisfying the following two conditions (i), (ii):*

(i) $\lambda_\nu > \sigma_\nu \geq 0, \ \lambda_\nu > \delta_\nu \geq 0,$ *and*

(ii) $\sigma_\nu \delta_\nu \equiv 1 \pmod{\lambda_\nu}, \quad \nu = 0, 1.$

Let s be a positive rational number such that

(iii) $-s + (\delta_0/\lambda_0 + \delta_1/\lambda_1)$ *is an integer.*

Then there exists uniquely a sequence of positive integers

$$(n_0, \ n_1, \ \ldots, \ n_l), \quad l \geq 1,$$

with the following properties:

(1) $n_0 = \lambda_0, \ \ n_l = \lambda_1,$

(2) $n_1 \equiv \sigma_0 \pmod{\lambda_0}, \quad n_{l-1} \equiv \sigma_1 \pmod{\lambda_1},$

(3) $n_{i-1} + n_{i+1} \equiv 0 \pmod{n_i}, \quad i = 1, \ldots, l-1,$

(4) $(n_{i-1} + n_{i+1})/n_i \geq 2, \quad i = 1, \ldots, l-1,$ *and*

(5) $\displaystyle\sum_{i=0}^{l-1} \frac{1}{n_i \, n_{i+1}} = s.$

Remark 5.1.

1. $\sigma_\nu = \delta_\nu = 0$ if and only if $\lambda_\nu = 1$. If $\lambda_\nu > 1$, then $\sigma_\nu, \delta_\nu \geq 1$.
2. The sequence $(n_0, n_1, \ldots, n_l)$ satisfies

$$\gcd(n_{i-1}, n_i) = 1, \quad i = 1, 2, \ldots, l,$$

Y. Matsumoto and J.M. Montesinos-Amilibia, *Pseudo-periodic Maps and Degeneration of Riemann Surfaces*, Lecture Notes in Mathematics 2030,
DOI 10.1007/978-3-642-22534-5_5, © Springer-Verlag Berlin Heidelberg 2011

because of properties (1),(2) and (3). *(NB.* $\gcd(\lambda_\nu, \sigma_\nu) = 1$.*)*

3. For a sequence with "length" $l = 1$, properties (3) and (4) are empty.

5.1 Proof of Theorem 5.1. (Uniqueness)

The uniqueness of $(n_0, n_1, \ldots, n_l)$ is contained in the following lemma:

Lemma 5.1. *Let* $(m_0, m_1, \ldots, m_k)$ *and* $(n_0, n_1, \ldots, n_l)$, $k \geq 1$, $l \geq 1$, *be sequences of positive integers satisfying*

(i) $m_0 = n_0$, $m_k = n_l$,

(ii) $m_1 \equiv n_1 \pmod{m_0}$, $m_{k-1} \equiv n_{l-1} \pmod{m_k}$,

(iii) $m_{i-1} + m_{i+1} \equiv 0 \pmod{m_i}$, $i = 1, 2, \ldots, k-1$,
$n_{j-1} + n_{j+1} \equiv 0 \pmod{n_j}$, $j = 1, 2, \ldots, l-1$,

(iv) $(m_{i-1} + m_{i+1})/m_i \geq 2$, $i = 1, 2, \ldots, k-1$,
$(n_{j-1} + n_{j+1})/n_j \geq 2$, $j = 1, 2, \ldots, l-1$, *and*

(v) $$\sum_{i=0}^{k-1} \frac{1}{m_i\, m_{i+1}} = \sum_{i=0}^{l-1} \frac{1}{n_j\, n_{j+1}}.$$

Then we have
$k = l$, *and* $m_i = n_i$, $i = 0, 1, \ldots, l$.

Claim (I). Let
$$(m_0, m_1, \ldots, m_k), \quad k \geq 1,$$
be a sequence of positive integers such that
$$(m_{i-1} + m_{i+1})/m_i \geq 2, \quad i = 1, 2, \ldots, k-1.$$

If $m_0 \geq m_k$, then $m_0 \geq m_1$.

Proof. If $k = 1$, the lemma is trivial. Let us assume $k \geq 2$. Suppose, on the contrary, that $m_0 < m_1$. The condition $(m_{i-1} + m_{i+1})/m_i \geq 2$ is rewritten as
$$m_i - m_{i-1} \leq m_{i+1} - m_i, \quad i = 1, \ldots, k-1.$$

Hence if $m_0 < m_1$, then $m_1 < m_2$ and inductively
$$m_0 < m_1 < m_2 < \cdots < m_k.$$

This contradicts the assumption $m_0 \geq m_k$. Thus $m_0 \geq m_1$. □

Proof (of Lemma 5.1.). By the symmetry of the statement, we may assume $m_0 \geq m_k$. By Claim (I), $m_0 \geq m_1$. By assumption (i), $n_0 = m_0$ and $n_l = m_k$, so $n_0 \geq n_l$. Then by Claim (I) again, $n_0 \geq n_1$.

Now we have

$$m_0 \geq m_1 > 0, \quad n_0 \geq n_1 > 0.$$

Then assumptions (i), (ii) imply $m_1 = n_1$, which together with assumption (iii) implies $m_2 \equiv n_2 \pmod{m_1}$. By assumption (v) and $m_0 = n_0$, $m_1 = n_1$, we have

$$\sum_{i=1}^{k-1} \frac{1}{m_i\, m_{i+1}} = \sum_{j=1}^{l-1} \frac{1}{n_j\, n_{j+1}}.$$

Thus the subsequences $(m_1, \ldots, m_k)$ and $(n_1, \ldots, n_l)$ satisfy the five conditions corresponding to (i)$\sim$(v). By induction on l, we have

$$k = l \quad \text{and} \quad (m_1, \ldots, m_k) = (n_1, \ldots, n_l).$$

We know already $m_0 = n_0$. Thus the lemma follows. □

The uniqueness of $(n_0, \ldots, n_l)$ is now established.

Before proving the existence, we will show a basic lemma.

Lemma 5.2. *Let*

$$(a_0, a_1, \ldots, a_u), \quad u \geq 1,$$

be a sequence of positive integers satisfying

(i) $\gcd(a_{i-1}, a_i) = 1, \quad i = 1, \ldots, u.$
(ii) $a_{i-1} + a_{i+1} \equiv 0 \pmod{a_i}, \quad a = 1, 2, \ldots, u-1.$

Then

1. $\displaystyle\sum_{i=0}^{u-1} \frac{1}{a_i\, a_{i+1}} \equiv \frac{d_0}{a_0} + \frac{e_u}{a_u} \pmod 1$, *where* d_0, e_u *are positive integers determined by*

$$0 < d_0 \leq a_0, \quad d_0\, a_1 \equiv 1 \pmod{a_0},$$

$$0 < e_u \leq a_u, \quad e_u\, a_{u-1} \equiv 1 \pmod{a_u}.$$

(NB. Here, for convenience, the inequalities are $0 < d_0 \leq a_0$, $0 < e_u \leq a_u$, not as usual $0 \leq d_0 < a_0$, $0 \leq e_u < a_u$; $d_0 = a_0$ if and only if $a_0 = 1$. Similarly for e_u).

2. *Moreover, if* $a_i \geq 2$, $i = 1, \ldots, u-1$, *then*

$$\sum_{i=0}^{u-1} \frac{1}{a_i\, a_{i+1}} = \frac{d_0}{a_0} + \frac{e_u}{a_u} - 1.$$

3. In particular, if $a_i \geq 2, i = 1, \ldots, u-1$*, and* $a_u = 1$*, then*

$$\sum_{i=0}^{u-1} \frac{1}{a_i\, a_{i+1}} = \frac{d_0}{a_0}.$$

Claim (J). Let a, a' be positive integers which are mutually coprime. Let d, e' be defined as follows

$$0 < d \leq a,\ \ da' \equiv 1 \pmod a$$
$$0 < e' \leq a',\ e'a \equiv 1 \pmod{a'}.$$

Then

$$\frac{1}{aa'} = \frac{d}{a} + \frac{e'}{a'} - 1.$$

Proof. Obviously $da' + e'a \equiv 1 \pmod a$ and $\equiv 1 \pmod{a'}$. Thus

$$da' + e'a = 1 + aa'z$$

for a certain $z \in \mathbf{Z}$, because $\gcd(a, a') = 1$. Hence

$$\frac{d}{a} + \frac{e'}{a'} = \frac{1}{aa'} + z.$$

By the assumption $0 < d \leq a,\ 0 < e' \leq a'$, we have $1/aa' < (d/a + e'/a') \leq 2$. Therefore, $z = 1$. □

Proof (of Lemma 5.2). Define two sequences of positive integers

$$(d_0,\ d_1,\ \ldots,\ d_{u-1}), \quad (e_1,\ e_2,\ \ldots,\ e_u)$$

as follows:

$$0 < d_i \leq a_i, \quad d_i\, a_{i+1} \equiv 1 \pmod{a_i}, \quad i = 0,\ 1,\ \ldots,\ u-1,$$
$$0 < e_i \leq a_i, \quad e_i\, a_{i-1} \equiv 1 \pmod{a_i}, \quad i = 1,\ 2,\ \ldots,\ u.$$

By the assumption (ii), $a_{i-1} + a_{i+1} \equiv 0 \pmod{a_i}$, we have $d_i + e_i \equiv 0 \pmod{a_i}$, i.e.

$$\frac{d_i}{a_i} + \frac{e_i}{a_i} = \text{an integer}.$$

By Claim (J),

$$\frac{1}{a_i\, a_{i+1}} = \frac{d_i}{a_i} + \frac{e_{i+1}}{a_{i+1}} - 1, \quad i = 0,\ \ldots,\ u-1.$$

Thus, adding up the equalities

$$\frac{1}{a_0 a_1} = \frac{d_0}{a_0} + \frac{e_1}{a_1} - 1$$

$$\frac{1}{a_1 a_2} = \frac{d_1}{a_1} + \frac{e_2}{a_2} - 1$$

$$\cdots\cdots\cdots\cdots\cdots$$

$$\frac{1}{a_{u-1} a_u} = \frac{d_{u-1}}{a_{u-1}} + \frac{e_u}{a_u} - 1$$

and taking into account that

$$\frac{e_1}{a_1} + \frac{d_1}{a_1}, \quad \frac{e_2}{a_2} + \frac{d_2}{a_2}, \quad \ldots$$

are integers, we get

$$\sum_{i=0}^{u-1} \frac{1}{a_i\, a_{i+1}} \equiv \frac{d_0}{a_0} + \frac{e_u}{a_u} \pmod 1.$$

This proves (1).

To prove (2), note that if $a_i \geq 2$, then $d_i/a_i + e_i/a_i = 1$ because in this case, $0 < d_i < a_i, 0 < e_i < a_i$ and $0 < (d_i/a_i + e_i/a_i) < 2$. Then the above addition gives

$$\sum_{i=0}^{u-1} \frac{1}{a_i\, a_{i+1}} = \frac{d_0}{a_0} + \frac{e_u}{a_u} - 1.$$

To prove (3), note that, if $a_u = 1$, then $e_u = 1$. Putting this in (2), we get

$$\sum_{i=0}^{u-1} \frac{1}{a_i\, a_{i+1}} = \frac{d_0}{a_0}.$$

□

5.2 Proof of Theorem 5.1. (Existence)

First we consider the general case where $\lambda_0, \lambda_1 > 1$ and hence $\sigma_0, \delta_0, \sigma_1, \delta_1 \geq 1$.

Let $(a_0, a_1, \ldots, a_u)$ be a sequence of positive integers satisfying

1. $a_0 = \lambda_0, \quad a_1 = \sigma_0,$
2. $a_0 > a_1 > \cdots > a_u = 1,$
3. $a_{i-1} + a_{i+1} \equiv 0 \pmod{a_i}, \quad i = 1, 2, \ldots, u-1.$

Such a sequence can be obtained by the Euclidean algorithm.

Likewise let $(b_0, b_1, \ldots, b_v)$ be a sequence of positive integers satisfying

1. $b_0 = \lambda_1, \ \ b_1 = \sigma_1$,
2. $b_0 > b_1 > \cdots > b_v = 1$,
3. $b_{j-1} + b_{j+1} \equiv 0 \pmod{b_j}, \ \ j = 1, 2, \ldots, v-1$.

By Lemma 5.2. (3),

$$\sum_{i=0}^{u-1} \frac{1}{a_i\, a_{i+1}} = \frac{\delta_0}{\lambda_0},$$

$$\sum_{j=0}^{v-1} \frac{1}{b_j\, b_{j+1}} = \frac{\delta_1}{\lambda_1}.$$

Let K denote $s - (\delta_0/\lambda_0 + \delta_1/\lambda_1)$, which is an integer by assumption (iii) of Theorem 5.1.

We have $K \geq -1$, because $s > 0$ and

$$0 < (\delta_0/\lambda_0 + \delta_1/\lambda_1) < 2.$$

Three cases are distinguished: $K > 0, \ \ K = 0, \ \ K = -1$.

Case I. $K > 0$. We define the required sequence $(n_0, n_1, \ldots, n_l)$ to be

$$(a_0,\ a_1,\ \ldots,\ a_u,\ \underbrace{1, \ldots, 1}_{K-1},\ b_v,\ \ldots,\ b_1,\ b_0).$$

This sequence has property (5) of Theorem 5.1:

$$\sum_{i=0}^{l-1} \frac{1}{n_i\, n_{i+1}} = \frac{\delta_0}{\lambda_0} + K + \frac{\delta_1}{\lambda_1} = s.$$

The other properties (1)~(4) are obvious by the construction.

Case II. $K = 0$. We define the sequence $(n_0, n_1, \ldots, n_l)$ to be

$$(a_0,\ a_1,\ \ldots,\ a_{u-1},\ 1,\ b_{v-1},\ \ldots,\ b_1,\ b_0).$$

Recall that $a_u = b_v = 1$.

Case III. $K = -1$. This is the most complicated case. We give several definitions. These definitions are motivated by the chain diagrams in [37] (Fig. 5.1):

Definition 5.1. An $*$-*sequence* γ of length $l \geq 1$ is a finite sequence

$$\gamma = (n_0,\ \varepsilon_1,\ n_1,\ \varepsilon_2,\ \ldots,\ \varepsilon_l,\ n_l)$$

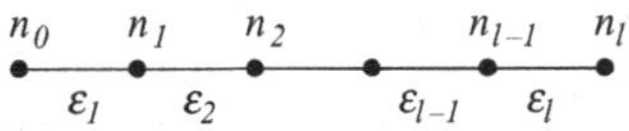

Fig. 5.1 Chain diagram

where the n_i's denote positive integers (called *the multiplicities of* γ) and ε_i is a *sign* $\varepsilon_i \in \{+, -\}$. Moreover, multiplicities and signs are required to satisfy the following '$*$-condition':

$$\varepsilon_i\, n_{i-1} + \varepsilon_{i+1}\, n_{i+1} \equiv 0 \pmod{n_i}, \quad i = 1,\, 2,\, \ldots,\, l-1. \qquad (*)$$

Of course, if $l = 1$, this condition is empty.

Remark 5.2. The $*$-condition implies that $\gcd(n_{i-1}, n_i)$ is independent of i.

Definition 5.2. The *s-number* $s(\gamma)$ of an $*$-sequence $\gamma = (n_0, \varepsilon_1, n_1, \varepsilon_2, \ldots, \varepsilon_l, n_l)$ is a rational number defined by

$$s(\gamma) = \sum_{i=0}^{l-1} \frac{\varepsilon_{i+1}}{n_i\, n_{i+1}}$$

Definition 5.3. Let

$$\gamma \;=\; (n_0,\, \varepsilon_1,\, n_1,\, \varepsilon_2,\, \ldots,\, \varepsilon_l,\, n_l)$$

be an $*$-sequence.

(i) Suppose there exists an $i \in \{1, 2, \ldots, l-1\}$ such that

$$\varepsilon_i\, n_{i-1} + \varepsilon_{i+1}\, n_{i+1} = \pm n_i.$$

Then the $*$-sequence

$$\gamma' = (n_0,\, \varepsilon_1,\, \ldots,\, \varepsilon_{i-1},\, n_{i-1},\, \varepsilon',\, n_{i+1},\, \varepsilon_{i+2},\, \ldots,\, n_l)$$

is said to be obtained from γ by *blowing down* the vertex of multiplicity n_i, where

$$\varepsilon' = sign\, [\varepsilon_i \varepsilon_{i+1}(\varepsilon_i n_{i-1} + \varepsilon_{i+1} n_{i+1})] = sign\, (\varepsilon_{i+1} n_{i-1} + \varepsilon_i n_{i+1}).$$

(ii) Suppose there exists an $i \in \{1, 2, \ldots, l-1\}$ such that

$$\varepsilon_i\, n_{i-1} + \varepsilon_{i+1}\, n_{i+1} = 0.$$

Then $n_{i-1} = n_{i+1}$ and $\varepsilon_i = -\varepsilon_{i+1}$. The $*$-sequence

$$\gamma'' = (n_0,\, \varepsilon_1,\, \ldots,\, n_{i-2},\, \varepsilon_{i-1},\, n',\, \varepsilon_{i+2},\, n_{i+2},\, \ldots, n_l)$$

is said to be obtained by *contracting* γ at the vertex of multiplicity n_i, where

$$n' = n_{i-1} = n_{i+1}.$$

It is not difficult to see that the operations of blowing down and contraction actually give an $*$-sequence. The operations do not change the s-numbers.

For simplicity, from now on, *a positive sign ε_i will be omitted* from the notation of an $*$-sequence.

For example, an $*$-sequence $(5, -, 3, +, 2, +, 1)$ will be denoted as $(5, -, 3, 2, 1)$. An $*$-sequence in which every sign is positive is called a *positive $*$-sequence*. Such a sequence will be denoted as $(n_0, n_1, \ldots, n_l)$ by the above convention.

Lemma 5.3. *If a positive $*$-sequence $\gamma = (n_0, n_1, \ldots, n_l), l \geq 1$, satisfies $n_0 > n_1 > \cdots > n_l$, then*

$$s(\gamma) < \frac{1}{n_l\,(n_{l-1} - n_l)}.$$

Proof. For each $i = 1, 2, \ldots, l-1$, there is a positive integer z_i such that

$$n_{i-1} + n_{i+1} = z_i\, n_i\,.$$

The assumption

$$n_0 > n_1 > \cdots > n_l$$

implies

$$z_i \geq 2.$$

Thus $n_{i-1} + n_{i+1} \geq 2\,n_i$, or equivalently,

$$n_{i-1} - n_i \geq n_i - n_{i+1}, \quad i = 1,\ 2,\ \ldots,\ l-1.$$

Setting $d = n_{l-1} - n_l$ (> 0), we inductively have

$$n_{l-j} \geq n_l + jd, \quad j = 0,\ 1,\ \ldots,\ l.$$

Then,

$$\begin{aligned} s(\gamma) &= \sum_{i=0}^{l-1} \frac{1}{n_i\, n_{i+1}} \\ &\leq \sum_{j=0}^{l-1} \frac{1}{(n_l + jd)\,[n_l + (j+1)\,d]} \\ &= \sum_{j=0}^{l-1} \left[\frac{1}{n_l + jd} - \frac{1}{n_l + (j+1)\,d}\right] \frac{1}{d} \\ &< \frac{1}{n_l\, d}. \end{aligned}$$

□

Lemma 5.4. *Let γ be an $*$-sequence*

$$(a_0,\ a_1,\ \dots,\ a_u,\ -,\ b_v,\ b_{v-1},\ \dots,\ b_0)$$

with a negative sign between the vertices of multiplicities a_u and b_v. Suppose that γ satisfies the following conditions, (i)~(iv):

(i) $u \geq 1,\ \ v \geq 1,$
(ii) $a_0 > a_1 > \cdots > a_u,\ \ b_0 > b_1 > \cdots > b_v,$
(iii) $a_{u-1} \geq b_v,\ b_{v-1} \geq a_u,$ *and*
(iv) $s(\gamma) > 0.$

Then (1) *or* (2) *occurs:*

1. *γ can be contracted to a positive $*$-sequence γ' which cannot be blown down anymore, or*
2. *γ can be blown down to an $*$-sequence γ' with one negative sign which satisfies the conditions corresponding to* (i)~(iv).

Moreover, in both cases, the terminal multiplicities a_0, b_0, of γ remain unchanged in γ'.

Remark 5.3. An $*$-sequence $(23, 7, 5, -, 2, 5, 23)$ shows that both (1) and (2) can occur simultaneously. In this case, by convention, we will always choose contraction.

Proof (of Lemma 5.4). Suppose $a_u = b_{v-1}$, then γ can be contracted at the vertex of multiplicity b_v. The resulting γ' is a positive $*$-sequence

$$(a_0,\ a_1,\ \dots,\ a_u(= b_{v-1}),\ b_{v-2},\ \dots,\ b_0).$$

Since

$$a_0 > a_1 > \cdots > a_u(= b_{v-1}) < b_{v-2} < \cdots < b_0,$$

this $*$-sequence cannot be further blown down.

Similarly, if $a_{u-1} = b_v, \gamma$ can be contracted to a positive $*$-sequence, which cannot be further blown down.

Henceforth we will assume $a_{u-1} > b_v$ and $b_{v-1} > a_u$. By the $*$-condition, we have

$$\begin{cases} a_{u-1} - b_v = \alpha a_u & (\exists\alpha : \text{a positive integer}), \\ b_{v-1} - a_u = \beta b_v & (\exists\beta : \text{a positive integer}). \end{cases} \qquad (**)$$

Case A. $\alpha = 1$ *and* $\beta = 1$: In this case, $a_{u-1} = a_u + b_v = b_{v-1}$. We will prove that $u = v = 1$ is impossible. Suppose on the contrary that $u = v = 1$. Then γ must be

$$(a_0,\ a_1,\ -,\ b_1,\ b_0)$$

with $a_0 = a_1 + b_1 = b_0$. We would have

$$s(\gamma) = \frac{1}{a_0\, a_1} - \frac{1}{a_1\, b_1} + \frac{1}{b_1\, b_0} = 0,$$

which contradicts condition (iv) on γ. Thus $u = v = 1$ is impossible.If $u \geq 2$, then, using $a_{u-1} - b_v = a_u$, we can blow down γ at the vertex of multiplicity a_u to obtain an $*$-sequence

$$\gamma' = (a_0,\ \ldots,\ a_{u-1},\ -,\ b_v,\ b_{v-1},\ \ldots,\ b_0),$$

in which $u - 1 \geq 1$. This γ' satisfies the conditions corresponding to (i)∼(iv). If $u \geq 2$, the argument is the same.

Case B. $\alpha = 1$ *and* $\beta \geq 2$: Since $a_{u-1} - b_v = a_u, \gamma$ can be blown down to an $*$-sequence

$$\gamma' = (a_0,\ \ldots,\ a_{u-1},\ -,\ b_v,\ b_{v-1},\ \ldots,\ b_0).$$

We will show that
(i)′ $u - 1 \geq 1$, and
(iii)′ $b_{v-1} > a_{u-1}$.
(The $*$-sequence trivially satisfies the conditions corresponding to (ii) and (iv)).

Proof (of (i)′). Let us suppose $u = 1$ and show that this leads to a contradiction. Equations $(**)$ with $\alpha = 1,\ \beta \geq 2,$ produce

$$\begin{cases} a_0 - b_v = a_1 \\ b_{v-1} - a_1 \geq 2b_v. \end{cases}$$

Hence it follows

$$b_v - b_{v-1} + a_0 \leq 0. \qquad (***)$$

γ' would have the form

$$(a_0,\ -,\ b_v,\ b_{v-1},\ \ldots,\ b_0).$$

Then

$$s(\gamma) = s(\gamma') = -\frac{1}{a_0\, b_v} + \sum_{i=0}^{v-1} \frac{1}{b_i\, b_{i+1}} < -\frac{1}{a_0\, b_v} + \frac{1}{b_v\,(b_{v-1} - b_v)}$$

(by Lemma 5.3)

$$= \frac{-b_{v-1} + b_v + a_0}{a_0 b_v (b_{v-1} - b_v)} \leq 0 \quad \text{by} \quad (***).$$

This contradicts condition (iv) on γ. Thus $u - 1 \geq 1$.

Proof (of (iii)′). From (**) with $\alpha = 1,\ \beta \geq 2,$ we have

$$\begin{cases} a_{u-1} = a_u + b_v, \\ b_{v-1} \geq a_u + 2b_v. \end{cases}$$

Thus $b_{v-1} > a_{u-1}$.

Case C. $\alpha \geq 2$ *and* $\beta = 1$: The argument is the same as in Case B.

Case D. $\alpha \geq 2$ *and* $\beta \geq 2$: From (**) with $\alpha \geq 2, \beta \geq 2$, we have

$$\begin{cases} a_{u-1} - b_v \geq 2a_u, \\ b_{v-1} - a_u \geq 2b_v. \end{cases}$$

Rewriting these inequalities, we obtain

$$\begin{cases} a_{u-1} - a_u \geq a_u + b_v, \\ b_{v-1} - b_v \geq a_u + b_v. \end{cases} \tag{$\star$}$$

Then

$$\begin{aligned} s(\gamma) &= \sum_{i=0}^{u-1} \frac{1}{a_i\, a_{i+1}} - \frac{1}{a_u\, b_v} + \sum_{j=0}^{v-1} \frac{1}{b_j\, b_{j+1}} \\ &< \frac{1}{a_u\,(a_{u-1} - a_u)} - \frac{1}{a_u\, b_v} + \frac{1}{b_v\,(b_{v-1} - b_v)} \qquad \text{(Lemma 5.3)} \\ &\leq \frac{1}{a_u\,(a_u + b_v)} - \frac{1}{a_u\, b_v} + \frac{1}{b_v\,(a_u + b_v)} \qquad \text{by } (\star) \\ &= 0. \end{aligned}$$

This contradicts condition (iv) on γ. Thus Case D is impossible.

Finally we can check that in every case discussed above, the terminal multiplicities a_0, b_0 remain unchanged. This completes the proof of Lemma 5.4. □

Let us continue the proof of Theorem 5.1 (existence) in Case (III); $K = -1$. Remember that

$$(a_0,\ a_1,\ \ldots,\ a_u)$$

is a positive $*$-sequence satisfying

$$a_0 = \lambda_0, \quad a_1 = \sigma_0, \quad a_0 > a_1 > \cdots > a_u = 1,$$

and

$$(b_0,\ b_1,\ \ldots,\ b_v)$$

is a similar sequence satisfying

$$b_0 = \lambda_1, \quad b_1 = \sigma_1, \quad b_0 > b_1 > \cdots > b_v = 1.$$

We define an $*$-sequence γ with one negative sign by

$$\gamma = (a_0,\ a_1,\ \dots,\ a_u,\ -,\ b_v,\ b_{v-1},\ \dots,\ b_0).$$

Since we are considering the general case where $\lambda_0 > \sigma_0 \geq 1$ and $\lambda_1 > \sigma_1 \geq 1$, we have $u \geq 1$ and $v \geq 1$ (assumption (i) of Lemma 5.4).

By the choice of

$$(a_0,\ a_1,\ \dots,\ a_u)$$

and

$$(b_0,\ b_1,\ \dots,\ b_v),$$

we have

$$a_0 > a_1 > \cdots > a_u$$

and

$$b_0 > b_1 > \cdots > b_v$$

(assumption (ii) of Lemma 5.4).

Since $a_u = b_v = 1$, we have $a_{u-1} > b_v$ and $b_{v-1} > a_u$ (assumption (iii)).

By Lemma 5.2 (3), we have

$$\sum_{i=0}^{u-1} \frac{1}{a_i\, a_{i+1}} = \frac{\delta_0}{\lambda_0}, \quad \text{and} \quad \sum_{j=0}^{v-1} \frac{1}{b_j\, b_{j+1}} = \frac{\delta_1}{\lambda_1}.$$

Thus

$$\begin{aligned} s(\gamma) &= \sum_{i=0}^{u-1} \frac{1}{a_i\, a_{i+1}} - \frac{1}{a_u\, b_v} + \sum_{j=0}^{v-1} \frac{1}{b_j\, b_{j+1}} \\ &= \frac{\delta_0}{\lambda_0} - 1 + \frac{\delta_1}{\lambda_1} = s > 0. \end{aligned}$$

This is assumption (iv) of Lemma 5.4.

Therefore, applying Lemma 5.4 to γ recursively, we will obtain a positive $*$-sequence

$$\gamma_0 = (n_0,\ n_1,\ \dots,\ n_l), \quad l \geq 1,$$

which can no longer be blown down.

This is the desired sequence; let us check the properties, (1)~(5).

Since the terminal multiplicities a_0, b_0 of γ remain unchanged in γ_0, we have

$$n_0 = a_0 = \lambda_0, \quad \text{and} \quad n_l = b_0 = \lambda_1 \quad (\text{ property (1) }).$$

Property (2) will be checked later.

Property (3) for γ_0 is obvious because γ_0 is an $*$-sequence.

Property (4) is true because γ_0 cannot be blown down.

Property (5) is true because neither contraction nor blowing down change the s-numbers, and γ has been chosen so that $s(\gamma) = s$.

Let us check property (2) for γ_0.

Let

$$\gamma_0' = (a_0', a_1', \ldots, a_h', -, b_k', b_{k-1}', \ldots, b_0'), \quad h \geq 1, k \geq 1,$$

be the $*$-sequence one step before γ_0; γ_0 is obtained from γ_0' by contraction. γ_0' has property (2) because $h \geq 1$, $k \geq 1$, $a_1' = a_1 = \sigma_0$ and $b_1' = b_1 = \sigma_1$.

Suppose γ_0 is obtained by contracting γ_0' at the vertex of the multiplicity a_h'. If $h \geq 2$, then a_1' remains in γ_0, and γ_0 has property (2). If $h = 1$, we have

$$\gamma_0 = (a_0' \ (= b_k'), b_{k-1}', \ldots, b_0').$$

By the $*$-condition on

$$\gamma_0' = (a_0', a_1', -, b_k', b_{k-1}', \ldots, b_0'),$$

we have $-a_1' + b_{k-1}' \equiv 0 \pmod{b_k'}$. Thus $b_{k-1}' \equiv a_1' \equiv \sigma_0 \pmod{a_0'}$ and

$$\gamma_0 = (a_0', b_{k-1}', \ldots, b_0')$$

has property (2).

If γ_0 is obtained by contracting γ_0' at the vertex of multiplicity b_k', the argument is the same.

This completes the proof of Theorem 5.1 in the general case where $\lambda_0, \lambda_1 > 1$.

Three special cases remain to be examined.

Special case (1). $\lambda_0 = 1$, $\lambda_1 > 1$. In this case $\sigma_0 = \delta_0 = 0$ and the integer $k = s - \delta_1/\lambda_1$ must be non-negative. Let $(b_0, b_1, \cdots, b_\nu)$ be an $*$-sequence satisfying

$$b_0 = \lambda_1, \quad b_1 = \sigma_1, \quad b_0 > b_1 > \cdots > b_\nu = 1.$$

Then the sequence

$$(\underbrace{1, \ldots, 1}_{k}, b_\nu, b_{\nu-1}, \ldots, b_0)$$

is the desired one.

Special case (2). $\lambda_0 > 1$, $\lambda_1 = 1$. The argument is the same as in Special case (1).

Special case (3). $\lambda_0 = \lambda_1 = 1$. In this case, $s = k =$ a positive integer. The desired sequence is

$$\underbrace{(1,\ 1,\ \ldots,\ 1)}_{k+1}$$

The proof of Theorem 5.1 is now completed. □

Remark 5.4. The sequence

$$(n_0,\ n_1,\ \ldots,\ n_l)$$

of Theorem 5.1 satisfies one of the following inequalities:

1. $n_0 > n_1 > \cdots > n_i = n_{i+1} = \cdots = n_j\ (= 1) < n_{j+1} < \cdots < n_l$, for some i, j with $0 < i < j < l$.
2. $n_0 > n_1 > \cdots > n_i < n_{i+1} < \cdots < n_l$, for some i with $0 < i < l$.
3. $n_0 > n_1 > \cdots > n_j = n_{j+1} = \cdots = n_l\ (= 1)$, for some $0 < j \le l$.
4. $n_0 > n_1 > \cdots > n_l\ (> 1)$.
5. $(1 =)\ n_0 = n_1 = \cdots = n_i < n_{i+1} < \cdots < n_l$, for some $0 \le i < l$.
6. $(1 <)\ n_0 < n_1 < \cdots < n_l$.
7. $n_0 = n_1 = \cdots = n_l\ (= 1)$.

Example 5.1. $(\lambda_0, \sigma_0, \delta_0) = (27, 17, 8)$, $(\lambda_1, \sigma_1, \delta_1) = (17, 10, 12)$.
Set

$$s = (\delta_0/\lambda_0 + \delta_1/\lambda_1 - 1) = 8/27 + 12/17 - 1 = 1/459 > 0.$$

Then

$$(a_0,\ a_1,\ a_2,\ \ldots,\ a_u) = (27,\ 17,\ 7,\ 4,\ 1), \quad u = 4$$

$$(b_0,\ b_1,\ b_2,\ \ldots,\ b_v) = (17,\ 10,\ 3,\ 2,\ 1), \quad v = 4.$$

$$\begin{aligned}
\gamma = \ & (27,\ 17,\ 7,\ 4,\ 1,\ -,\ 1,\ 2,\ 3,\ 10,\ 17) \\
& \downarrow 1 \ \text{(meaning "blowing down at 1")} \\
& (27,\ 17,\ 7,\ 4,\ 1,\ -,\ 2,\ 3,\ 10,\ 17) \\
& \downarrow 2 \\
& (27,\ 17,\ 7,\ 4,\ 1,\ -,\ 3,\ 10,\ 17) \\
& \downarrow 1 \\
& (27,\ 17,\ 7,\ 4,\ -,\ 3,\ 10,\ 17) \\
& \downarrow 4 \\
& (27,\ 17,\ 7,\ -,\ 3,\ 10,\ 17) \\
& \downarrow 3 \\
& (27,\ 17,\ 7,\ -,\ 10,\ 17) \\
& \downarrow 7 \\
& (27,\ 17,\ -,\ 10,\ 17) \\
& \downarrow \text{contracting } 10 \\
\gamma_0 = \ & (27,\ 17).
\end{aligned}$$

Chapter 6
Conjugacy Invariants

Let $f : \Sigma_g \to \Sigma_g$ be a pseudo-periodic map of negative twist. According to Corollary 4.5, the isomorphism class of the minimal quotient

$$\pi : \Sigma_g \to S[f]$$

and, in particular, the numerical homeomorphism type of $S[f]$ are conjugacy invariants of $[f] \in \mathscr{M}_g$. However, the converse is not true: the minimal quotient does not necessarily determine the conjugacy class of $[f]$. (Nielsen [50, Sect. 15] incorrectly claims that this converse is true (compare Theorem 13.4 of [22] where this same claim is repeated)).

Example 6.1. In Fig. 6.1, $f_1|\Sigma_2 - C_1 \simeq$ 180°rotation around the (vertical) axis; $s(C_1) = -2$. The minimal quotient space $S[f_1]$ is depicted in Fig. 6.2.

Example 6.2. In Fig. 6.3, $f_2|\Sigma_2 - C_1 \simeq$ 180°rotation around the (horizontal) axis; $s(C_1) = -2$. The minimal quotient space $S[f_2]$ is depicted in Fig. 6.4.

The minimal quotient spaces $S[f_1]$ and $S[f_2]$ are numerically homeomorphic, but $[f_1]$ and $[f_2]$ are not conjugate, because the amphidrome curve in Example 6.1 is null-homologous while it is not the case in Example 6.2. In fact, the minimal quotients $\pi_1 : \Sigma_g \to S[f_1]$ and $\pi_2 : \Sigma_g \to S[f_2]$ are not isomorphic. (*Proof.* Let P_0 be the closed part (Chap. 4) in the irreducible component of genus 1, which is contained in both minimal quotient spaces. Then $\pi_1^{-1}(P_0)$ has two connected components, while $\pi_2^{-1}(P_0)$ is connected).

Remark 6.1. The pseudo-periodic maps f_1 and f_2 are the topological monodromies of the singular fibers 2-$2\mathrm{I}_{0-0}$ and 3-II^*_{1-0} in Namikawa-Ueno's classification [49], respectively. They pointed out that these singular fibers have the same configuration but different monodromies.

Y. Matsumoto and J.M. Montesinos-Amilibia, *Pseudo-periodic Maps and Degeneration of Riemann Surfaces*, Lecture Notes in Mathematics 2030,
DOI 10.1007/978-3-642-22534-5_6, © Springer-Verlag Berlin Heidelberg 2011

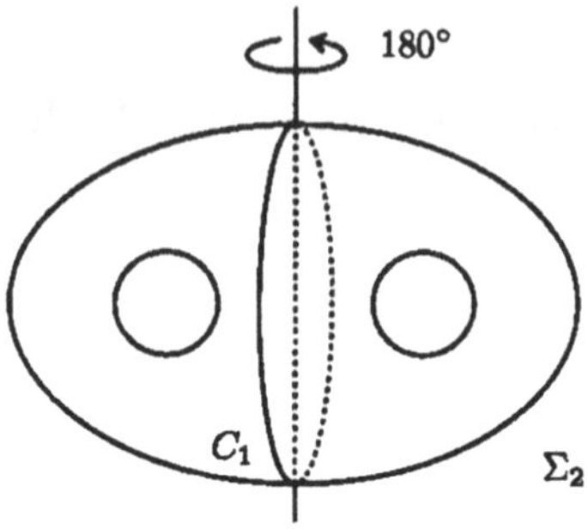

Fig. 6.1 Example 6.1: $f_1|\Sigma_2 - C_1$ is a "180°-rotation" and $s(C_1) = -2$

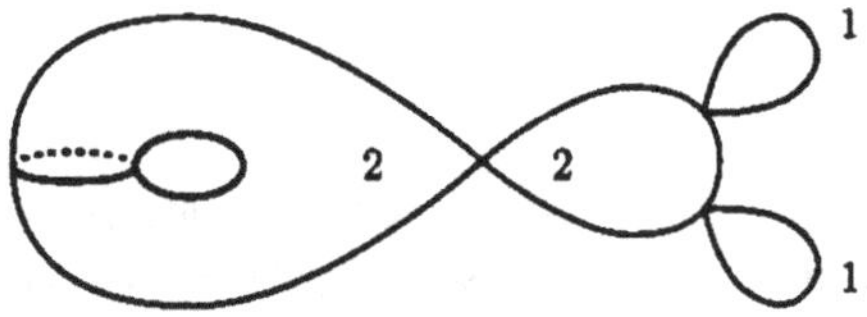

Fig. 6.2 The minimal quotient of f_1 of Example 6.1

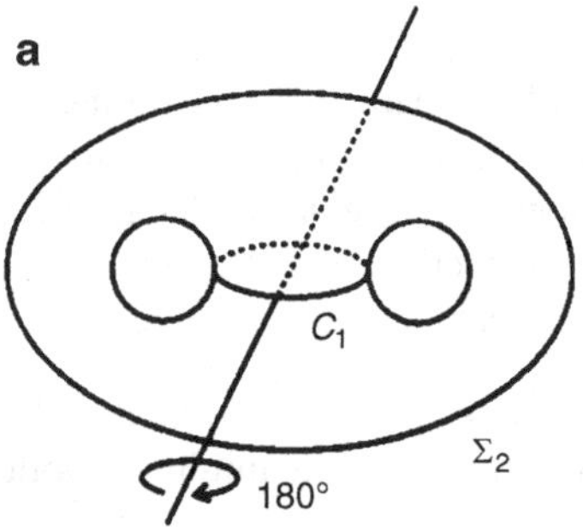

Fig. 6.3 Example 6.2: $f_2|\Sigma_2 - C_1$ is a "180°-rotation" and $s(C_1) = -2$

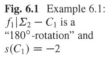

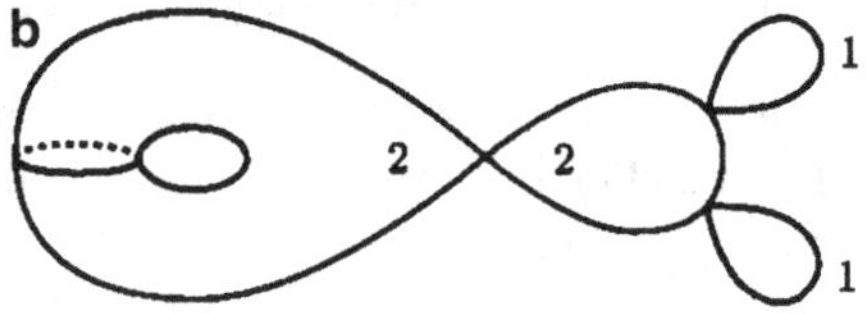

Fig. 6.4 The minimal quotient of f_2 of Example 6.2

Example 6.3. In Fig. 6.5, $f_3|\Sigma_6 - \bigcup_{i=1}^{5} C_i \simeq 1/5$ turn rotation; $s(C_i) = -1$, $i = 1, 2, ..., 5$. The minimal quotient space $S[f_3]$ is depicted in Fig. 6.6.

Example 6.4. In Fig. 6.7, $f_4|\Sigma_6 - \bigcup_{i=1}^{5} C_i \simeq 2/5$ turn rotation; $s(C_i) = -1$, $i = 1, 2, ..., 5$. The minimal quotient space $S[f_4]$ is depicted in Fig. 6.8

The minimal quotients

$$\pi_3 : \Sigma_6 \to S[f_3]$$

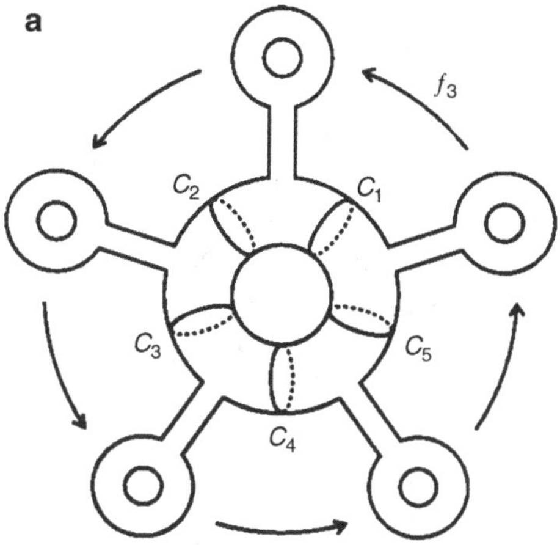

Fig. 6.5 Example 6.3: $f_3|\Sigma_6 - \bigcup_{i=1}^{5} C_i$ is a $1/5$-turn and $s(C_i) = -1$

Fig. 6.6 The minimal quotient of f_3 of Example 6.3

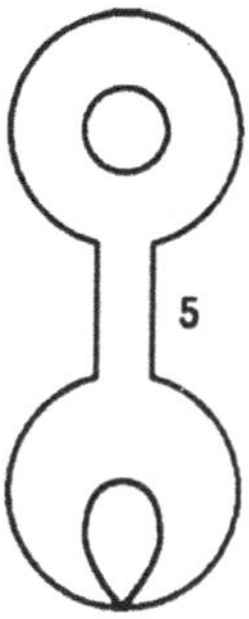

and

$$\pi_4 : \Sigma_6 \to S[f_4]$$

are *isomorphic* as pinched coverings, but $[f_3]$ and $[f_4]$ are not conjugate, because the actions of f_3 and f_4 on the "partition graphs" are different. (See Corollary 6.1).

Remark 6.2. f_4 is not conjugate to $(f_3)^2$; the minimal quotient space of $(f_3)^2$, $S[(f_3)^2]$, is depicted in Fig. 6.9.

The purpose of this chapter is to show that the action of a pseudo-periodic map f on its partition graph gives the information lacking in the minimal quotient and that that action, together with the minimal quotient, determines the conjugacy class of $[f]$ in $\mathscr{M}_g$. (Theorems 6.1 and 6.3).

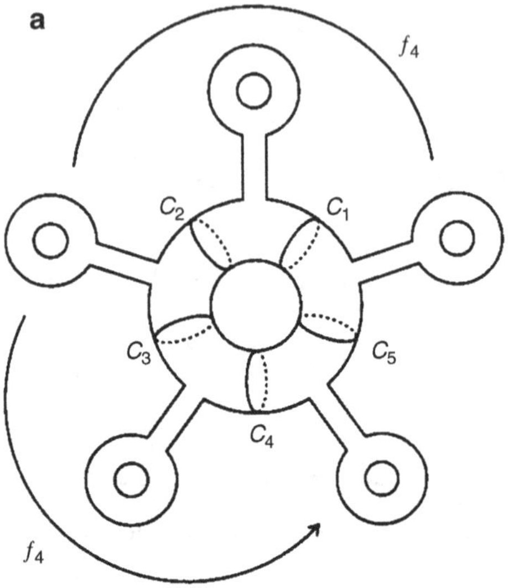

Fig. 6.7 Example 6.4: $f_4|\Sigma_6 - \bigcup_{i=1}^{5} C_i$ is a $2/5$-turn and $s(C_i) = -1$

Fig. 6.8 The minimal quotient of f_4 of Example 6.4

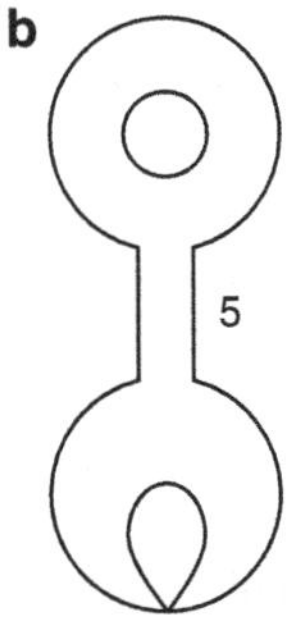

Fig. 6.9 The minimal quotient of $(f_3)^2$

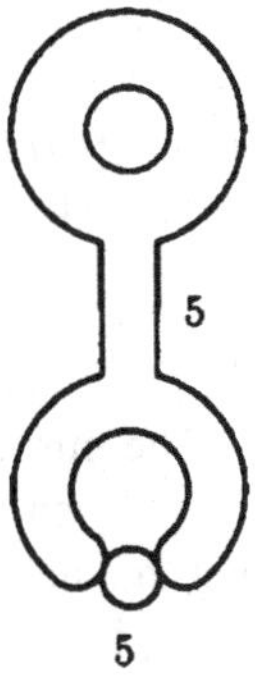

6.1 Partition Graphs

Definition 6.1. A *graph* is a 1-dimensional finite cell complex. A 1-cell is called an *edge*, a 0-cell a *vertex*. There may be loops. There may be loops sharing the same vertex. Also there may be a component consisting of one vertex. A homeomorphism $\varphi : X \to X'$ between graphs is called an *isomorphism* if it preserves the graph structure.

Definition 6.2. Let $f : \Sigma_g \to \Sigma_g$ be a pseudo-periodic map of negative twist, $\{C_i\}_{i=1}^r$ the precise system of cut curves subordinate to f (Chap. 1). The *partition graph* $X[f]$ *associated with* f is the graph whose vertices are in one-to-one correspondence to the connected components of $\Sigma_g - \bigcup_{i=1}^r C_i$ and whose edges are in one-to-one correspondence to the cut curves $\{C_i\}_{i=1}^r$. An edge $e(C_i)$ joins vertices $v(b)$ and $v(b')$ if and only if C_i is in the adherence of the connected component b and at the same time in that of the connected component b'. It may happen that $b = b'$, in which case $e(C_i)$ is a loop.

By Nielsen's theorem (Theorem 2.2), the isomorphism class of the partition graph $X[f]$ is a conjugacy invariant of $[f]$.

Definition 6.3. The *collapsing map*

$$\xi : \Sigma_g \to X[f]$$

is defined as follows: Let $\{A_i\}_{i=1}^r$ be the annular neighborhood system of $\{C_i\}_{i=1}^r$. Each A_i can be identified with $[0,1] \times S^1$ through a parametrization

$$\phi_i : [0,1] \times S^1 \to A_i.$$

The map

$$\xi : \Sigma_g \to X[f]$$

shrinks each connected component of $\Sigma_g - \bigcup_{i=1}^r Int(A_i)$ to the corresponding vertex and projects each annulus $A_i = [0,1] \times S^1$ onto its first factor $[0,1]$ which is identified with the corresponding edge. We assume that ξ maps a cut curve C_i to the middle point of the edge $e(C_i)$.

It is not difficult to prove that the collapsing map

$$\xi : \Sigma_g \to X[f]$$

is well-defined up to isotopy of Σ_g. (Cf. Nielsen's theorem, Theorem 2.2.)

The standard form $\overline{f}$ ("fitted" to $\{A_i\}_{i=1}^r$) projects to an isomorphism

$$\varphi_{[f]} : X[f] \to X[f]$$

such that $\varphi_{[f]}|e$ is linear on each edge e.

Of course, we have the commutative diagram:

$$\begin{array}{ccc} \Sigma_g & \xrightarrow{\overline{f}} & \Sigma_g \\ {\scriptstyle \xi}\downarrow & & \downarrow{\scriptstyle \xi} \\ X[f] & \xrightarrow[\varphi_{[f]}]{} & X[f]. \end{array}$$

Since $X[f]$ is a finite graph, the map $\varphi_{[f]}$ is a periodic map.

Definition 6.4. $\varphi_{[f]} : X[f] \to X[f]$ is said to be the *periodic map induced by* f. f acts on $X[f]$ through $\varphi_{[f]}$.

Lemma 6.1. *Let* $f_1 : \Sigma_g^{(1)} \to \Sigma_g^{(1)}$ *and* $f_2 : \Sigma_g^{(2)} \to \Sigma_g^{(2)}$ *be pseudo-periodic maps of negative twist. Suppose there exists a homeomorphism*

$$h : \Sigma_g^{(1)} \to \Sigma_g^{(2)}$$

such that $f_1 \simeq h^{-1} f_2 h$ *(homotopic), then there exist a homeomorphism*

$$h' : \Sigma_g^{(1)} \to \Sigma_g^{(2)}$$

and an isomorphism

$$\Phi : X[f_1] \to X[f_2]$$

such that

(i) h' *is isotopic to* h*, and*
(ii) the following diagram commutes:

$$\begin{array}{ccc} \Sigma_g^{(1)} & \xrightarrow{h'} & \Sigma_g^{(2)} \\ {\scriptstyle \xi_1}\downarrow & & \downarrow{\scriptstyle \xi_2} \\ X[f_1] & \xrightarrow{\Phi} & X[f_2] \\ {\scriptstyle \varphi_{[f_1]}}\downarrow & & \downarrow{\scriptstyle \varphi_{[f_2]}} \\ X[f_1] & \xrightarrow{\Phi} & X[f_2]. \end{array}$$

Proof. The lemma follows from the uniqueness of the standard form (Theorem 2.1). □

Corollary 6.1. *The (equivariant) isomorphism class of* $(X[f], \varphi_{[f]})$ *is a conjugacy invariant of* $[f] \in \mathcal{M}_g$.

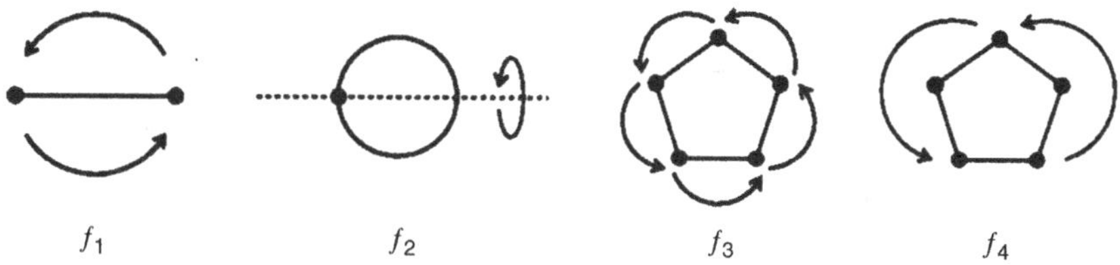

Fig. 6.10 Partition graphs of f_i's in Examples $6.i,\ i = 1, 2, 3, 4$

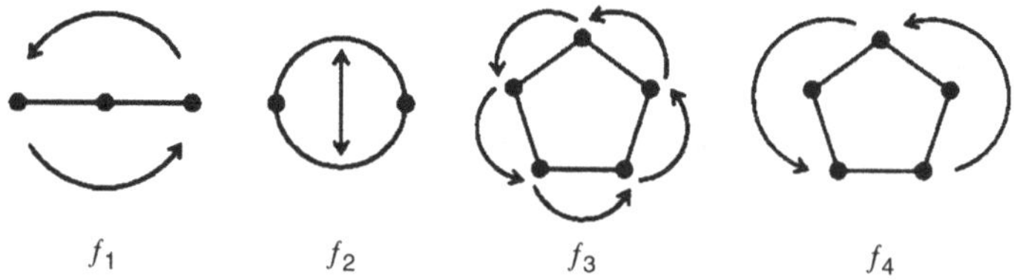

Fig. 6.11 Refined partition graphs of f_i's in Examples $6.i,\ i = 1, 2, 3, 4$

Examples. Partition graphs and the actions of the f's corresponding to Examples 6.1 through 6.4. See Fig. 6.10.

Definition 6.5. Let $f : \Sigma_g \to \Sigma_g$ be a pseudo-periodic map of negative twist. The *refined partition graph* $\overline{X}[f]$ *associated with* f is obtained from $X[f]$ by subdividing each edge $e(C_i)$ corresponding to an *amphidrome* curve C_i into two edges by the middle point. The middle point becomes a new vertex $v(C_i)$ of $\overline{X}[f]$. The collapsing map $\xi : \Sigma_g \to \overline{X}[f]$ is defined to be the same map $\xi : \Sigma_g \to X[f]$ as before. (Thus ξ maps an amphidrome curve C_i to the vertex $v(C_i)$.)

Examples. Refined partition graphs corresponding to Examples 6.1 through 6.4 (Fig. 6.11).

6.2 Weighted Graphs

An advantage of the refined partition graph $\overline{X}[f]$ is that the periodic action of $\varphi_{[f]}$ has no multiple points in the interior of an edge. Thus if we take the quotient space $Y[f] = \overline{X}[f]/\varphi_{[f]}$, the projection map

$$\rho : \overline{X}[f] \longrightarrow Y[f]$$

sends each edge of $\overline{X}[f]$ homeomorphically onto an edge of $Y[f]$.

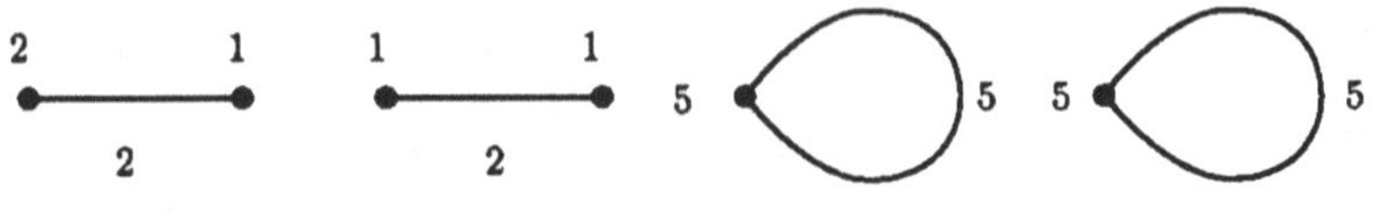

Fig. 6.12 Weighted graphs

Definition 6.6. A graph Y is *weighted* if each vertex (and each edge) carries a positive integer called the *weight*. An isomorphism $\Psi : Y \to Y'$ between weighted graphs is called a *weighted isomorphism* if Ψ preserves the weights.

The quotient graph $Y[f]$ becomes a weighted graph as follows: the weight of a vertex v (resp. an edge e) is defined to be $\sharp\rho^{-1}(v)$ (resp. $\sharp\rho^{-1}(e)$).

Definition 6.7. $Y[f]$ is called the *weighted graph associated with* $f : \Sigma_g \to \Sigma_g$.

Examples. Weighted graphs corresponding to Examples 6.1 through 6.4 (Fig. 6.12):

The weighted graph $Y[f]$ has another meaning. It serves as the *decomposition diagram* of the base chorizo space of the minimal quotient $\pi : \Sigma_g \to S[f]$, i.e. the vertices of $Y[f]$ are in one-to-one correspondence to the bodies, and their edges are in one-to-one correspondence to the arches of $S[f]$ (Cf. Chap. 4). More precisely, we have the following

Lemma 6.2. *There exists a map*

$$\eta : S[f] \to Y[f]$$

satisfying the following conditions:

(i) *η shrinks a body of $S[f]$ to a vertex and maps an arch onto an edge of $Y[f]$.*
(ii) *η induces a one-to-one correspondence between the set of the bodies (resp. arches) and the set of the vertices (resp. edges).*
(iii) *For a body BDY_ν of $S[f]$, the weight of the vertex $\eta(BDY_\nu)$ is equal to the number of the connected components of $\pi^{-1}(BDY_\nu)$, and for an arch $ARCH_\mu$ of $S[f]$, the weight of the edge $\eta(ARCH_\mu)$ is equal to the number of the annuli in $\pi^{-1}(ARCH_\mu)$.*
(iv) *The following diagram "almost" commutes in the sense that ξ can be deformed on amphidrome annuli so that the diagram commutes with the modified ξ':*

$$\begin{array}{ccc} \Sigma_g & \xrightarrow{\xi'} & \overline{X}[f] \\ {\scriptstyle\pi}\downarrow & & \downarrow{\scriptstyle\rho} \\ S[f] & \xrightarrow[\eta]{} & Y[f]. \end{array}$$

Proof. Let $\overline{f}$ be the standard form of f. Let BDY_ν be a body of $S[f]$. If BDY_ν is an ordinary body, then $\pi^{-1}(BDY_\nu)$ consists of a certain number of compact surfaces with negative Euler characteristic (Lemma 4.4). Pick up one component B_j, map it by ξ to a vertex of $\overline{X}[f]$, and project it down to a vertex of $Y[f]$. The image $\eta(BDY_\nu)$ is defined to be this vertex. Since $\overline{f}$ permutes the components of $\pi^{-1}(BDY_\nu)$ cyclically (Proposition 4.2), the vertex $\eta(BDY_\nu)$ is well-defined, independently of the choice of the connected component of $\pi^{-1}(BDY_\nu)$. The weight of $\eta(BDY_\nu)$ is equal to the number of the vertices of $\overline{X}[f]$ which are over the vertex $\eta(BDY_\nu)$, but by the definition of

$$\xi : \Sigma_g \to \overline{X}[f],$$

this number is equal to the number of the connected components of $\pi^{-1}(BDY_\nu)$.

If BDY_ν is a special body, $BDY(m)$, with multiplicity $2m$ (see Fig. 4.2), then $\pi^{-1}(BDY(m))$ consists of m annuli

$$B'_1,\ B'_2,\ \ldots,\ B'_m,$$

(Corollary 4.3). Let $ARCH_0$ be the (special) arch attached to $BDY(m)$. Then

$$\pi^{-1}(ARCH_0 \cup BDY(m))$$

consists of m amphidrome annuli,

$$A_1,\ A_2,\ \ldots,\ A_m,$$

and B'_i is the middle one-third of A_i. Pick up one annulus A_i.

$$\xi : \Sigma_g \to \overline{X}[f]$$

maps the center-line C_i of A_i to the vertex $v(C_i)$ of $\overline{X}[f]$. The image $\eta(BDY(m))$ is defined to be the vertex $\rho(v(C_i)) \in Y[f]$, which is well-defined because $\overline{f}$ permutes

$$\{A_1,\ A_2,\ \ldots,\ A_m\}$$

cyclically. The weight of the vertex $\eta(BDY(m))$ is certainly m.

Deform

$$\xi : \Sigma_g \to \overline{X}[f]$$

on $\bigcup_{i=1}^m A_i$ so that the resulting ξ' sends each middle one-third B'_i of A_i to the vertex $v(C_i)$, then we have $\rho\xi' = \eta\pi$ on $\pi^{-1}(BDY(m))$.

Let $ARCH_\mu$ be an (ordinary or special) arch of $S[f]$. The preimage $\pi^{-1}(ARCH_\mu)$ consists of a certain number of annuli, which are permuted by $\overline{f}$ cyclically (Lemma 4.2). Pick up an annulus A'_i in $\pi^{-1}(ARCH_\mu)$. If $ARCH_\mu$ is ordinary, A'_i is mapped by ξ onto an edge $e(C_i)$ of $\overline{X}[f]$. Project down $e(C_i)$ onto an edge of $Y[f]$ to get

$$\eta : ARCH_\mu \to \rho(e(C_i)).$$

Clearly we have $\rho\xi = \eta\pi$ on A'_i.

If $ARCH_\mu$ is special, then A_i' is an outer one-third of an annulus A_i (Lemma 4.5). The annulus A_i' is mapped by ξ' (which has been deformed so that it sends B_i' to the vertex $v(C_i)$) onto an edge $e'(C_i)$ of $\overline{X}[f]$, where $e'(C_i)$ is a half of the edge $e(C_i)$ of $X[f]$. Project down $e'(C_i)$ onto an edge of $Y[f]$ to obtain

$$\eta : ARCH_\mu \to \rho(e'(C_i)).$$

Then we have $\rho\xi' = \eta\pi$ on A_i'.

Putting together the above partially defined maps, we get the desired "collapsing" map

$$\eta : S[f] \to Y[f].$$

The properties (i)~(iv) are now easy to check. □

Definition 6.8. The map

$$\eta : S[f] \to Y[f]$$

is called the *collapsing map* of the chorizo space $S[f]$.

Corollary 6.2. *Let*

$$\eta : S[f] \to Y[f]$$

be the collapsing map.

(i) *The weight of the edge $\eta(ARCH_\mu)$ is the* gcd *of the successive multiplicities on $ARCH_\mu$.*
(ii) *Let BDY_ν be a body of $S[f]$. If the core part P_0 of BDY_ν is of genus* 0, *then the weight of the vertex $\eta(BDY_\nu)$ is equal to* $\gcd(m_0, m_1, \ldots, m_k)$, *where m_0, $m_1, \ldots, m_k$ have the same meaning as in Proposition 4.2.*
(iii) *If P_0 is of genus ≥ 1, the weight of the vertex $\eta(BDY_\nu)$ is a common divisor of $m_0, m_1, \ldots, m_k$.*
(iv) *The weight of a vertex $v \in Y[f]$ is a common divisor of the weights of the edges emanating from v.*

The proof is straightforward by Lemmmas 6.2, 4.2, and Proposition 4.2. Assertion (iv) follows from (i)~(iii).

Remark 6.3. The weighted graph $Y[f]$ generalizes the graph considered in [18,48].

Theorem 6.1. *Let*

$$f_1 : \Sigma_g^{(1)} \to \Sigma_g^{(1)} \quad \text{and} \quad f_2 : \Sigma_g^{(2)} \to \Sigma_g^{(2)}$$

be pseudo-periodic maps of negative twist, both in superstandard form. Let

$$\pi_1 : \Sigma_g^{(1)} \to S[f_1] \quad \text{and} \quad \pi_2 : \Sigma_g^{(2)} \to S[f_2]$$

be the respective minimal quotients. Then there exists a homeomorphism

$$h : \Sigma_g^{(1)} \to \Sigma_g^{(2)}$$

such that $f_1 = h^{-1} f_2 h$ *if and only if there exist a numerical homeomorphism*

$$H : S[f_1] \to S[f_2],$$

a weighted isomorphism

$$\Psi : Y[f_1] \to Y[f_2]$$

and an isomorphism

$$\Phi : \overline{X}[f_1] \to \overline{X}[f_2]$$

such that the diagram commutes:

$$\begin{array}{ccccccc} S[f_1] & \xrightarrow{\eta_1} & Y[f_1] & \xleftarrow{\rho_1} & \overline{X}[f_1] & \xleftarrow{\varphi_{[f_1]}} & \overline{X}[f_1] \\ {\scriptstyle H}\downarrow & & {\scriptstyle \Psi}\downarrow & & {\scriptstyle \Phi}\downarrow & & {\scriptstyle \Phi}\downarrow \\ S[f_2] & \xrightarrow[\eta_2]{} & Y[f_2] & \xleftarrow[\rho_2]{} & \overline{X}[f_2] & \xleftarrow[\varphi_{[f_2]}]{} & \overline{X}[f_2]. \end{array}$$

Remark 6.4. The homeomorphism $h : \Sigma_g^{(1)} \to \Sigma_g^{(2)}$ obtained above does not necessarily project to the given numerical homeomorphism H.

Proof (of Theorem 6.1). The "only if" part is easy to prove. The "if"part is essential. We may assume $H : S[f_1] \to S[f_2]$ preserves the systems of closed nodal neighborhoods.

Let $BDY_\nu^{(1)}$ be a body in $S[f_1]$, $BDY_\nu^{(2)} = H(BDY_\nu^{(1)})$ its image. Since

$$\Psi\eta_1(BDY_\nu^{(1)}) = \eta_2 H(BDY_\nu^{(1)}) = \eta_2(BDY_\nu^{(2)}),$$

the number of connected components of $\pi_1^{-1}(BDY_\nu^{(1)})$ and that of $\pi_2^{-1}(BDY_\nu^{(2)})$ are equal (see Lemma 6.2(iii)). Then by Proposition 4.3, there exist a homeomorphism

$$h : \pi_1^{-1}(BDY_\nu^{(1)}) \to \pi_2^{-1}(BDY_\nu^{(2)})$$

and a numerical homeomorphism

$$H' : BDY_\nu^{(1)} \to BDY_\nu^{(2)}$$

such that

1. $f_1 = (h)^{-1} f_2 h$ on $\pi_1^{-1}(BDY_\nu^{(1)})$,
2. $\pi_2 h = H'\pi_1$ on $\pi_1^{-1}(BDY_\nu^{(1)})$, and
3. $H'|\partial BDY_\nu^{(1)} = H|\partial BDY_\nu^{(1)}$.

Note that the connected components of $\pi_1^{-1}(BDY_\nu^{(1)})$ and $\pi_2^{-1}(BDY_\nu^{(2)})$ are permuted cyclically by f_1 and f_2, respectively, and that the equality (1) implies that $h : \pi_1^{-1}(BDY_\nu^{(1)}) \to \pi_2^{-1}(BDY_\nu^{(2)})$ preserves these cyclic orders. Thus, replacing h by $h(f_1)^l$ with a certain integer l, we may assume the following equality holds:

4. $\xi_2 h = \Phi \xi_1 \quad \text{on} \quad \pi^{-1}(BDY_\nu^{(1)})$.

(Here

$$\xi_1 : \Sigma_g^{(1)} \to \overline{X}[f_1] \quad \text{and} \quad \xi_2 : \Sigma_g^{(2)} \to \overline{X}[f_2]$$

denote the respective collapsing maps).

Let $ARCH_\mu^{(1)}$ be an arch in $S[f_1]$,

$$ARCH_\mu^{(2)} = H(ARCH_\mu^{(1)})$$

its image. Since

$$H : S[f_1] \to S[f_2]$$

preserves the multiplicities, $\pi_1^{-1}(ARCH_\mu^{(1)})$ and $\pi_2^{-1}(ARCH_\mu^{(2)})$ consists of the same number of annuli (Lemma 4.2), and by Corollary 4.1, there exists a homeomorphism

$$h : \pi_1^{-1}(ARCH_\mu^{(1)}) \to \pi_2^{-1}(ARCH_\mu^{(2)})$$

such that

5. $f_1 = (h)^{-1} f_2 h \quad \text{on} \quad \pi_1^{-1}(ARCH_\mu^{(1)})$, and
6. $\pi_2 h = H \pi_1 \quad \text{on} \quad \pi_1^{-1}(ARCH_\mu^{(1)})$.

By the same reason as in case of BDY_ν, h can be assumed to satisfy

7. $\xi_2 h = \Phi \xi_1 \quad \text{on} \quad \pi_1^{-1}(ARCH_\mu^{(1)})$.

Now our task it to paste together these homeomorphisms

$$\{h \mid \pi_1^{-1}(BDY_\nu^{(1)})\}_\nu \quad \text{and} \quad \{h \mid \pi_1^{-1}(ARCH_\mu^{(1)})\}_\mu$$

to produce a homeomorphism $h : \Sigma_g^{(1)} \to \Sigma_g^{(2)}$. If it is done, the pasted h will satisfy $f_1 = (h)^{-1} f_2 h$ by (1) and (5).

To see if this pasting is possible, let us fix our attention at an edge e of $Y[f_1]$ and the terminal vertices v_0 and v_1. (It may happen that $v_0 = v_1$, when e is a loop). Let $BDY_\nu^{(1)}$ be the body in $S[f_1]$ such that $\eta(BDY_\nu^{(1)}) = v_\nu$ $(\nu = 0, 1)$, $ARCH^{(1)}$ the arch in $S[f_1]$ such that $\eta(ARCH^{(1)}) = e$. $BDY_\nu^{(1)}$ has a certain number of boundary components, $\nu = 0, 1$. We assume $ARCH^{(1)}$ is attached to a boundary curve $\Gamma_0^{(1)}$ of $BDY_0^{(1)}$ and to a boundary curve $\Gamma_1^{(1)}$ of $BDY_1^{(1)}$.

Take a connected component $B_0^{(1)}$ of $\pi_1^{-1}(BDY_0^{(1)})$. Over the boundary curve $\Gamma_0^{(1)}$, there are a certain number (say c_0) of boundary curves of $B_0^{(1)}$. Let them be

$$\tilde{\Gamma}_{0,1}^{(1)},\ \tilde{\Gamma}_{0,2}^{(1)},\ \ldots,\ \tilde{\Gamma}_{0,c_0}^{(1)}.$$

Consider one of them, say $\tilde{\Gamma}_{0,1}^{(1)}$. There is an annulus $A^{(1)}$ over $ARCH^{(1)}$ such that $\tilde{\Gamma}_{0,1}^{(1)}$ is one of its two boundary curves. Suppose the other boundary curve of $A^{(1)}$ is $\tilde{\Gamma}_{1,1}^{(1)}$, where

$$\tilde{\Gamma}_{1,1}^{(1)},\ \tilde{\Gamma}_{1,2}^{(1)},\ \ldots,\ \tilde{\Gamma}_{1,c_1}^{(1)}$$

are the boundary curves of a certain connected component, say $B_1^{(1)}$, of $\pi_1^{-1}(BDY_1^{(1)})$, which lie over $\tilde{\Gamma}_1^{(1)}$.

Set $B_0^{(2)} = h(B_0^{(1)})$, $B_1^{(2)} = h(B_1^{(1)})$ and $A^{(2)} = h(A^{(1)})$. Also set $\tilde{\Gamma}_{0,\alpha}^{(2)} = h(\tilde{\Gamma}_{0,\alpha}^{(1)})$ $(\alpha = 1, \ldots, c_0)$ and $\tilde{\Gamma}_{1,\beta}^{(2)} = h(\tilde{\Gamma}_{1,\beta}^{(1)})$ $(\beta = 1, \ldots, c_1)$.

By (4) and (7), it is clear that one of the two boundary curves of $A^{(2)}$ is among

$$\{\tilde{\Gamma}_{0,1}^{(2)},\ \tilde{\Gamma}_{0,2}^{(2)},\ \ldots,\ \tilde{\Gamma}_{0,c_0}^{(2)}\}$$

and the other is among

$$\{\tilde{\Gamma}_{1,1}^{(2)},\ \tilde{\Gamma}_{1,2}^{(2)},\ \ldots,\ \tilde{\Gamma}_{1,c_1}^{(2)}\},$$

but to paste together the homeomorphisms $h|B_0^{(1)}$, $h|B_1^{(1)}$ and $h|A^{(1)}$, we need the equality

$$\partial A^{(2)} = \tilde{\Gamma}_{0,1}^{(2)} \cup \tilde{\Gamma}_{1,1}^{(2)}$$

to hold. However, we cannot expect this in general. The difficulty will be overcome by the following lemma:

Lemma 6.3. *Let Σ be a compact connected surface whose Euler characteristic is negative, $f : \Sigma \to \Sigma$ a periodic map of order m_0 without multiple points. Let $M = \Sigma/f$ be the quotient space with the projection $\pi : \Sigma \to M$. Take a boundary component Γ_1 of M, and let*

$$\tilde{\Gamma}_{1,1},\ \tilde{\Gamma}_{1,2},\ \ldots,\ \tilde{\Gamma}_{1,\mu}$$

be the totality of the preimages of Γ_1 under π. Suppose they are indexed so that $f(\tilde{\Gamma}_{1,\alpha}) = \tilde{\Gamma}_{1,\alpha+1}$, $\alpha = 1, \ldots, \mu - 1$, and $f(\tilde{\Gamma}_{1,\mu}) = \tilde{\Gamma}_{1,1}$.

Then there exists a homeomorphism $j : \Sigma \to \Sigma$ such that

(A) $jf = fj$
(B) $j(\tilde{\Gamma}_{1,\alpha}) = \tilde{\Gamma}_{1,\alpha+1}$, $\alpha = 1, \ldots, \mu - 1$, *and* $j(\tilde{\Gamma}_{1,\mu}) = \tilde{\Gamma}_{1,1}$,
(C) $j|(\partial\Sigma - \bigcup_{\alpha=1}^{\mu} \tilde{\Gamma}_{1,\alpha}) = id$.

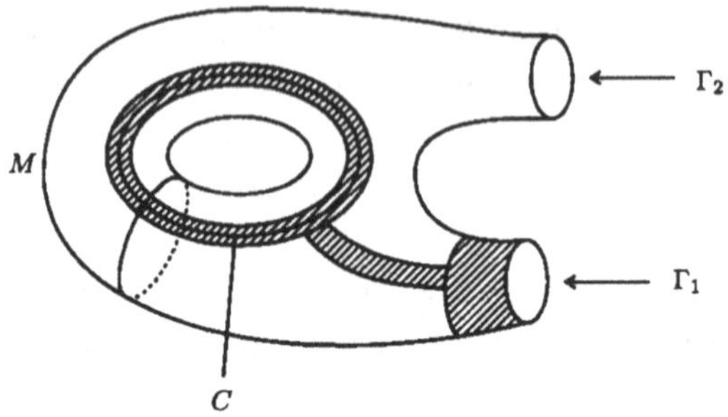

Fig. 6.13 A band connected sum $N(C)\natural N(\Gamma_1)$ is identified with Δ_2

Proof. Suppose M has k boundary components, $\Gamma_1, \Gamma_2, \ldots, \Gamma_k$, each oriented from the inside. Let

$$\omega : H_1(M) \to \mathbf{Z}/m_0$$

be the monodromy exponent (see Theorem 1.3, Proposition 3.1, and [51, Sect. 2]). Set $\omega(\Gamma_i) = m_i, i = 1, 2, \ldots, k$.

Case 1. *M has genus* ≥ 1. In this case, there exists a simple closed (oriented) curve C in M such that $\omega(C) = 1$, ([51, Sect. 5]). Take an annular neighborhood $N(C)$ of C and a collar neighborhood $N(\Gamma_1)$ of Γ_1, then make a band connected sum $N(C)\natural N(\Gamma_1)$ of $N(C)$ and $N(\Gamma_1)$ inside M. (Fig. 6.13). Identifying $N(C)\natural N(\Gamma_1)$ with a disk with two holes Δ_2, we perform a full Dehn-twist inside Δ_2 about a curve parallel to the boundary $\partial_0 \Delta_2$ not homologous to Γ_1 nor to C, then rotate the boundary curve $\partial_C \Delta_2$ parallel to C, $(-m_1 - 1)$ times. (See Fig. 6.14 where $m_1 = 2$.) This gives a homeomorphism

$$J : \Delta_2 \to \Delta_2$$

with $J|\partial\Delta_2 = id$. We extend

$$J : \Delta_2 \to \Delta_2$$

to a homeomorphism

$$J : M \to M$$

by defining $J|(M - Int\Delta_2)$ to be the identity. Let c be an arc in Δ_2 connecting $\partial_C \Delta_2$ (the boundary curve parallel to C) and the boundary curve $\partial_0 \Delta_2$. Then since

$$\omega(J(c) - c) = \omega(\Gamma_1) + \omega(C) - (m_1 + 1)\omega(C) = 0, \quad \text{Fig. 6.14,}$$

we can lift J to a homeomorphism

$$j : \Sigma \to \Sigma$$

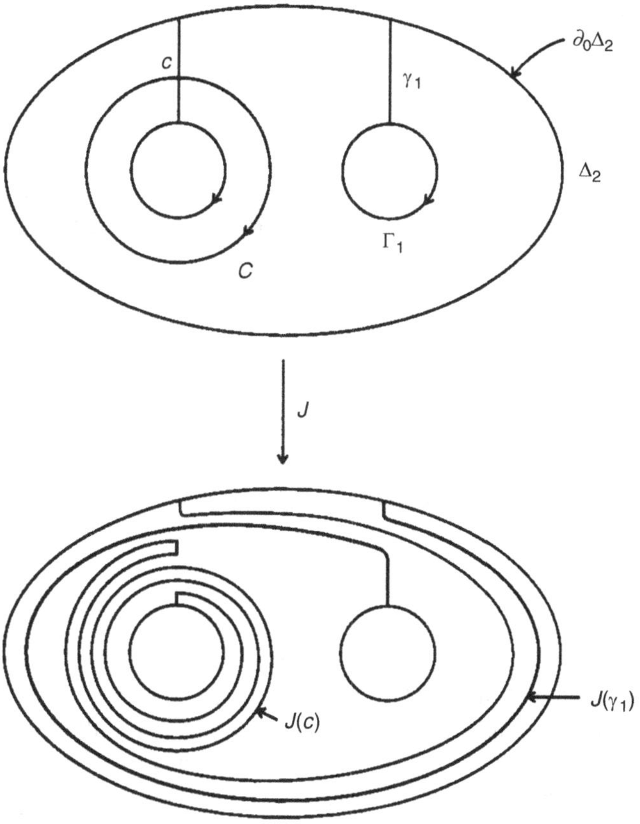

Fig. 6.14 Perform a full Dehn twist about a curve parallel to $\partial_0 \Delta_2$, then rotate $\partial_C \Delta_2$ $(-m_1 - 1)$ times

such that $j|(\Sigma - \pi^{-1}(Int(\Delta_2)) = id$. By the definition of the monodromy exponent $\omega : H_1(M) \to \mathbf{Z}/m_0$, there are $\gcd(m_0, m_1)$ boundary curves of Σ over the boundary curve Γ_1 of M. [51, Sect. 2]. Thus, in particular, the number μ of the boundary curves over Γ_1 is equal to $\gcd(m_0, m_1)$. Let γ_1 be an arc in Δ_2 connecting Γ_1 and $\partial_0 \Delta_2$. Since the arc γ_1 is mapped to $J(\gamma_1)$ satisfying

$$\omega(J(\gamma_1) - \gamma_1) = \omega(\Gamma_1) + \omega(C) = m_1 + 1, \quad \text{Fig. 6.14},$$

the lifted homeomorphism

$$j : \Sigma \to \Sigma$$

sends $\tilde{\Gamma}_{1,\alpha}$ to $\tilde{\Gamma}_{1,\alpha+1}$ $(\alpha = 1, \ldots, \mu - 1)$ and $\tilde{\Gamma}_{1,\mu}$ to $\tilde{\Gamma}_{1,1}$.The homeomorphism j satisfies $jf = fj$ because j is a lift of a homeomorphism of M and j has fixed points in Σ. This completes the proof in Case 1.

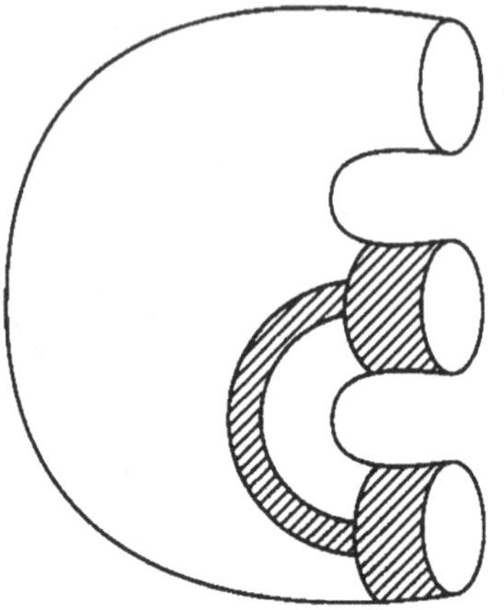

Fig. 6.15 A band connected sum $N(\Gamma_1)\natural N(\Gamma_2)$ is identified with Δ_2

Case 2. *M is of genus* 0. Since $\chi(M) = (\chi(\Sigma))/m_0 < 0$, M has more than two boundary curves

$$\Gamma_1, \ \Gamma_2, \ \ldots, \ \Gamma_k, \quad k \geq 3.$$

Take collar neighborhoods $N(\Gamma_1)$ and $N(\Gamma_2)$ of Γ_1 and Γ_2, respectively, and make a band connected sum $N(\Gamma_1)\natural N(\Gamma_2)$ inside M. (See Fig. 6.15). Identifying $N(\Gamma_1)\natural N(\Gamma_2)$ with a disk with two holes Δ_2, we perform a full Dehn twist about a curve parallel to the boundary curve $\partial_0\Delta_2$ other than Γ_1 or Γ_2. (See Fig. 6.16). This gives us a homeomorphism $J_2 : \Delta_2 \to \Delta_2$ which extends by the identity to a homeomorphism $J_2 : M \to M$. We lift J_2 to a homeomorphism $j_2 : \Sigma \to \Sigma$ such that $j_2|(\Sigma - \pi^{-1}(Int\Delta_2)) = id$. Let γ_i be an arc in Δ_2 which connects $\partial_0\Delta_2$ and $\Gamma_i, i = 1, 2$. Then

$$\omega(J_2(\gamma_1) - \gamma_1) = m_1 + m_2\omega(J_2(\gamma_2) - \gamma_2) = m_1 + m_2, \quad \text{Fig. 6.16.}$$

Thus j_2 maps each boundary curve $\tilde{\Gamma}_{1,\alpha}$ (over Γ_1) to $\tilde{\Gamma}_{1,\alpha+m_2}$ (the indices are considered to be in $\mathbf{Z}/\mu = \mathbf{Z}/\gcd(m_0, m_1)$, and each boundary curve $\tilde{\Gamma}_{2,\beta}$ (over Γ_2) to $\tilde{\Gamma}_{2,\beta+m_1}$ (the indices are considered to be in $\mathbf{Z}/\gcd(m_0, m_2)$). Apply this homeomorphism j_2 m_2 times, and set $j_2' = (j_2)^{m_2}$. Then j_2' maps $\tilde{\Gamma}_{1,\alpha}$ to $\tilde{\Gamma}_{1,\alpha+(m_2)^2}$, and $\tilde{\Gamma}_{2,\beta}$ to $\tilde{\Gamma}_{2,\beta+m_2m_1} = \tilde{\Gamma}_{2,\beta}$. A certain amount of rotation about $\tilde{\Gamma}_{2,\beta}$ might be caused by j_2', but it can be rotated back to the identity, equivariantly with respect to f. The modified j_2' sends the boundary curve (over Γ_1) of index α to the boundary curve (over Γ_1) of index $\alpha + (m_2)^2$, $\alpha = 1, \ldots, \mu$, and restricts to the identity of $\partial\Sigma - \bigcup_{\alpha=1}^{\mu} \tilde{\Gamma}_{1,\alpha}$. We do the same construction using Γ_1 and Γ_i $(i = 2, 3, \ldots, k)$. Then obtain a homeomorphism j_i', for each $i = 2, 3, \ldots, k$, which sends the boundary curve (over Γ_1) of index α to the one of index $\alpha + (m_i)^2$, and restricts to the identity on $\partial\Sigma - \bigcup_{\alpha=1}^{\mu} \tilde{\Gamma}_{1,\alpha}$. The (j_i')'s commute with f by the same reason as in Case 1. Since Σ is connected, we have:

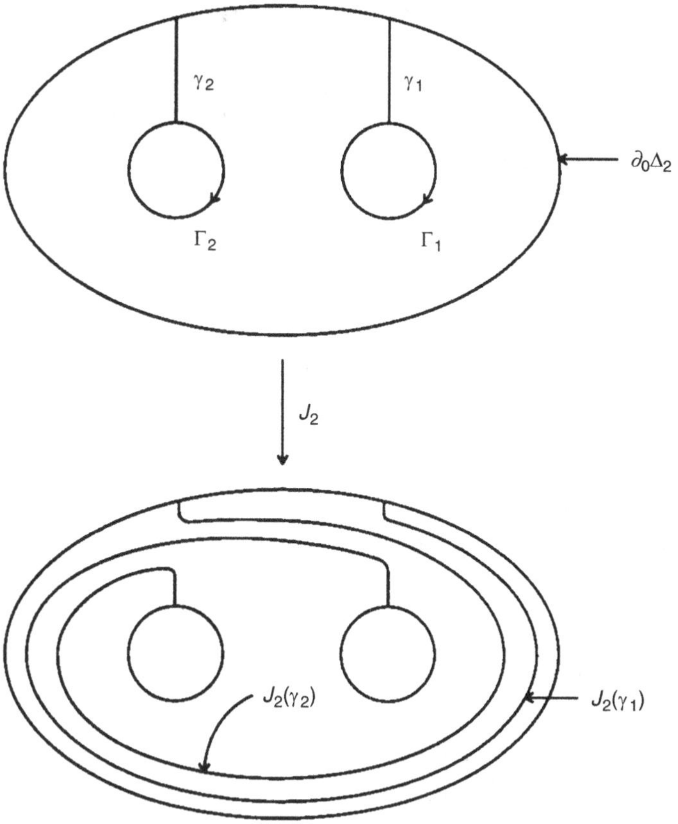

Fig. 6.16 Perform a full Dehn twist about a curve parallel to $\partial_0 \Delta_2$

$$\gcd(m_0, m_1, m_2, \ldots, m_k) = 1, \quad [51, \text{Sect. } 4(4.4)],$$

which implies

$$\gcd(m_0, m_1, (m_2)^2, \ldots, (m_k)^2) = 1.$$

Let l_0, l_1, ..., l_k be the integers such that

$$l_0 m_0 + l_1 m_1 + l_2 (m_2)^2 + \cdots + l_k (m_k)^2 = 1.$$

Then the homeomorphism $j : \Sigma \to \Sigma$, defined by

$$j = (j_2')^{l_2} (j_3')^{l_3} \cdots (j_k')^{l_k},$$

satisfies the conditions (A), (B), (C), of Lemma 6.3. This completes the proof of Lemma 6.3 in Case 2. □

6.3 Completion of the Proof of Theorem 6.1

Proof. Let us return to the situation before Lemma 6.3. Let μ_0 (resp. μ_1) be the number of the connected components of $\pi_2^{-1}(BDY_0^{(2)})$ (resp. $\pi_2^{-1}(BDY_1^{(2)})$), and let P_0 (resp. P_1) be the core part of $BDY_0^{(2)}$ (resp. $BDY_1^{(2)}$). Set $\tilde{P}_0 = \pi_2^{-1}(P_0) \cap B_0^{(2)}$, and $\tilde{P}_1 = \pi_2^{-1}(P_1) \cap B_1^{(2)}$.

Recall that $\tilde{P}_0$ has c_0 boundary curves

$$\tilde{\Gamma}_{0,1}^{(2)},\ \tilde{\Gamma}_{0,2}^{(2)},\ \ldots,\ \tilde{\Gamma}_{0,c_0}^{(2)}$$

over the boundary curve $\Gamma_0^{(2)} = H(\Gamma_0^{(1)})$ of $BDY_0^{(2)}$, and likewise $\tilde{P}_1$ has c_1 boundary curves

$$\tilde{\Gamma}_{1,1}^{(2)},\ \tilde{\Gamma}_{1,2}^{(2)},\ \ldots,\ \tilde{\Gamma}_{1,c_1}^{(2)}$$

over the boundary curve $\Gamma_1^{(2)} = H(\Gamma_1^{(1)})$ of $BDY_1^{(2)}$.

Applying Lemma 6.3 to the periodic map

$$(f_2)^{\mu_0} : \tilde{P}_0 \to \tilde{P}_0,$$

we obtain a homeomorphism $j_0 : \tilde{P}_0 \to \tilde{P}_0$ such that

(A) $j_0(f_2)^{\mu_0} = (f_2)^{\mu_0} j_0$,
(B) $j_0(\tilde{\Gamma}_{0,\alpha}^{(2)}) = \tilde{\Gamma}_{0,\alpha+1}^{(2)} \quad (\alpha \in \mathbf{Z}/c_0)$, and
(C) $j_0|(\partial\tilde{P}_0 - \bigcup_{\alpha=1}^{c_0} \tilde{\Gamma}_{0,\alpha}^{(2)}) = id$.

Extend $j_0 : \tilde{P}_0 \to \tilde{P}_0$ by the identity to

$$j_0 : B_0^{(2)} \to B_0^{(2)}$$

having the same properties (A), (B), (C). Then replacing the homeomorphism

$$h : B_0^{(1)} \to B_0^{(2)}$$

by $(j_0)^l h$, with a certain integer l, we can cyclically change the positions of the $\tilde{\Gamma}_{0,\alpha}^{(2)}$'s (without moving

$$\partial B_0^{(2)} - \bigcup_{\alpha=1}^{c_0} \tilde{\Gamma}_{0,\alpha}^{(2)}$$

at all), and adjust so that one of the two boundary components of $A^{(2)}$ $(= h(A^{(1)}))$ is attached to $\tilde{\Gamma}_{0,1}^{(2)}$ $(= h(\tilde{\Gamma}_{0,1}^{(1)}))$.

By the same argument, we can find a homeomorphism

$$j_1 : \tilde{P}_1 \to \tilde{P}_1$$

with the similar properties (A), (B), (C) and extend it to

$$j_1 : B_1^{(2)} \to B_1^{(2)}.$$

Again changing cyclically the positions of the $\tilde{\Gamma}_{1,\alpha}^{(2)}$'s using j_1, we can adjust so that the other boundary component of $A^{(2)}$ is attached to $\tilde{\Gamma}_{1,1}^{(2)}$ $(= h(\tilde{\Gamma}_{1,1}^{(1)}))$.

Note that the modified $h|B_0^{(1)}$ and $h|B_1^{(1)}$ still satisfy

8. $(f_2)^{\mu_0}(h|B_0^{(1)}) = (h|B_0^{(1)})(f_1)^{\mu_0}$ and
9. $(f_2)^{\mu_1}(h|B_1^{(1)}) = (h|B_1^{(1)})(f_1)^{\mu_1}$

because of property (A).

Now we can paste the homeomorphisms $h|B_0^{(1)}$, $h|B_1^{(1)}$ and $h|A^{(1)}$. We must be a little careful, because on the intersection $\tilde{\Gamma}_{0,1}^{(1)} = B_0^{(1)} \cap A^{(1)}$ the homeomorphisms $h|B_0^{(1)}$ and $h|A^{(1)}$ might be different by a certain amount of rotation. The same thing can be said on the intersection $\tilde{\Gamma}_{1,1}^{(1)} = B_1^{(1)} \cap A^{(1)}$. Thus, in general, we must compose

$$h|A^{(1)} : A^{(1)} \to A^{(2)}$$

with a certain linear twist

$$l : A^{(2)} \to A^{(2)}$$

so that the composed lh coincides with $h|B_0^{(1)}$ (resp. $h|B_1^{(1)}$) on the boundary curve $\tilde{\Gamma}_{0,1}^{(1)}$ (resp. $\tilde{\Gamma}_{1,1}^{(1)}$). If this is done, denoting the composition

$$lh : A^{(1)} \to A^{(2)}$$

again by

$$h : A^{(1)} \to A^{(2)},$$

we will have the pasted homeomorphism

$$h : B_0^{(1)} \cup A^{(1)} \cup B_1^{(1)} \to B_0^{(2)} \cup A^{(2)} \cup B_1^{(2)}.$$

Let m be the number of the annuli in $\pi_1^{-1}(ARCH^{(1)})$. We extend the above homeomorphism (equivariantly) to

$$h_\alpha : (f_1)^{\alpha-1}(B_0^{(1)} \cup A^{(1)} \cup B_1^{(1)}) \to (f_2)^{\alpha-1}(B_0^{(2)} \cup A^{(2)} \cup B_1^{(2)}), \quad \alpha = 1, \dots, m,$$

by the formula

10. $h_\alpha = (f_2)^{\alpha-1}(h)(f_1)^{1-\alpha}$.

Note that if $(f_1)^{\alpha-1}(B_0^{(1)}) = B_0^{(1)}$, then $\alpha - 1$ is a multiple of μ_0, say $k\mu_0$, and we have

$$(f_2)^{\alpha-1}(h|B_0^{(1)})(f_1)^{1-\alpha} = (f_2)^{k\mu_0}(h|B_0^{(1)})(f_1)^{-k\mu_0} = h|B_0^{(1)}$$

by (8). Similarly, if $(f_1)^{\alpha-1}(B_1^{(1)}) = B_1^{(1)}$, then

$$(f_2)^{\alpha-1}(h|B_1^{(1)})(f_2)^{1-\alpha} = h|B_1^{(1)}$$

by (9).

Therefore, formula (10) gives well-defined homeomorphisms h_α, $\alpha = 1, \ldots, m$, and since

$$\pi_1^{-1}(BDY_0^{(1)} \cup ARCH^{(1)} \cup BDY_1^{(1)}) = \bigcup_{\alpha=1}^{m}(f_1)^{\alpha-1}(B_0^{(1)} \cup A^{(1)} \cup B_1^{(1)}),$$

and

$$\pi_2^{-1}(BDY_0^{(2)} \cup ARCH^{(2)} \cup BDY_1^{(2)}) = \bigcup_{\alpha=1}^{m}(f_2)^{\alpha-1}(B_0^{(2)} \cup A^{(2)} \cup B_1^{(2)}),$$

we have a homeomorphism,

$$h = \bigcup_{\alpha=1}^{m} h_\alpha : \pi_1^{-1}(BDY_0^{(1)} \cup ARCH^{(1)} \cup BDY_1^{(1)}) \to \pi_2^{-1}(BDY_0^{(2)} \cup ARCH^{(2)} \cup BDY_1^{(2)}),$$

such that $f_2 h = h f_1$. (This last equality is verified using Corollary 4.1.)

So far we have confined ourselves to the part over an edge e of $Y[f_1]$. But the same argument can be carried out for the part over every edge e, and we can paste up the component homeomorphisms

$$\{h|\pi_1^{-1}(BDY_\nu^{(1)})\}_\nu \text{ and } \{h|\pi^{-1}(ARCH_\mu^{(1)})\}_\mu$$

(after proper modifications as we did above) to produce a homeomorphism

$$h : \Sigma_g^{(1)} \to \Sigma_g^{(2)}$$

such that $f_1 = (h)^{-1} f_2 h$.

This completes the proof of Theorem 6.1. □

6.4 Weighted Cohomology

We would like to reformulate Theorem 6.1 in terms of the "weighted cohomology" of a weighted graph which will be introduced below.

Let Y be a weighted graph.

Definition 6.9. A 0-*cochain* c^0 is an operation which assigns to each vertex v an element of a cyclic group $\mathbf{Z}/W(v)$, where $W(v)$ denotes the weight attached to v. A 1-*cochain* c^1 is an operation which assigns to each oriented edge $\overrightarrow{e}$ an element of $\mathbf{Z}/\gcd(W(v), W(w))$, where $\partial\overrightarrow{e} = v - w$, such that $c^1(-\overrightarrow{e}) = -c^1(\overrightarrow{e})$.

Let $C^0(Y)$ and $C^1(Y)$ denote the groups of 0-cochains and 1-cochains, respectively. Then the *coboundary operator* $\delta : C^0(Y) \to C^1(Y)$ is defined as follows:

$$(\delta(c^0))(\overrightarrow{e}) = c^0(v) - c^0(w) \in \mathbf{Z}/\gcd(W(v), W(w)),$$

where $\partial\overrightarrow{e} = v - w$.

Definition 6.10. The *weighted cohomology group* $H^1_W(Y)$ or Y is defined by

$$H^1_W(Y) = C^1(Y)/\delta C^0(Y).$$

Let X be a graph, $\varphi : X \to X$ an isomorphism such that $\varphi|e$ is linear for each edge e. Since we are considering only finite graphs, φ is a periodic map. We assume the following condition:

$$\varphi \text{ has no multiple points in the interior of an edge.}$$

Then we have the quotient graph $Y = X/\varphi$. Let $\rho : X \to Y$ be the projection. By condition $(*)$, each edge of X is mapped onto an edge of Y homeomorphically. Y naturally becomes a weighted graph: the weight of a vertex υ is $\sharp\rho^{-1}(\upsilon)$, and the weight of an edge e is $\sharp\rho^{-1}(e)$.

In what follows we will construct a cohomology class in $H^1_W(Y)$ which classifies the (equivariant) isomorphism class of (X, φ).

Let υ be a vertex of Y. The preimage $\rho^{-1}(\upsilon)$ contains $W(\upsilon)$ vertices, $\tilde{\upsilon}_1, \tilde{\upsilon}_2, \ldots,$ $\tilde{\upsilon}_{W(\upsilon)}$. We will always assume that the indices are elements of $\mathbf{Z}/W(\upsilon)$ and satisfy

$$\varphi(\tilde{\upsilon}_\alpha) = \tilde{\upsilon}_{\alpha+1}, \quad \alpha = 1, 2, \ldots, W(\upsilon).$$

Suppose all the vertices in $\rho^{-1}(\upsilon)$ have been indexed as above, for every vertex υ of Y.

Let e be an oriented edge of Y with $\partial e = v - w$. Let $\tilde{e}$ be an edge of X which is a lift of e. Suppose $\partial\tilde{e} = \tilde{\upsilon}_\alpha - \tilde{w}_\beta$, where $\alpha \in \mathbf{Z}/W(\upsilon)$ and $\beta \in \mathbf{Z}/W(w)$. Then we define a 1-cochain $c^1 \in C^1(Y)$ by setting

$$c^1(e) = \alpha - \beta \in \mathbf{Z}/\gcd(W(\upsilon), W(w)).$$

The value $\alpha-\beta$ in $\mathbf{Z}/\gcd(W(\upsilon), W(w))$ is independent of the choice of a lift of e. To see this, let $\tilde{e}'$ be another lift of e. Suppose $\partial\tilde{e}' = \tilde{\upsilon}_\sigma - \tilde{w}_\tau$, where $\sigma \in \mathbf{Z}/W(\upsilon)$ and $\tau \in \mathbf{Z}/W(w)$. Since $\rho(\tilde{e}) = \rho(\tilde{e}') = e$, there exists a power of φ, say φ^l, such that $\varphi^l(\tilde{e}) = \tilde{e}'$, which implies

$$\varphi^l(\tilde{\upsilon}_\alpha) = \tilde{\upsilon}_\sigma \quad \text{and} \quad \varphi^l(\tilde{w}_\beta) = \tilde{w}_\tau.$$

By our assumption on the manner of indexing, we have

$$\alpha + l = \sigma \in \mathbf{Z}/W(\upsilon), \quad \text{and} \quad \beta + l = \tau \in \mathbf{Z}/W(w).$$

Thus $\alpha - \beta = \sigma - \tau$ in $\mathbf{Z}/\gcd(W(\upsilon), W(w))$ as asserted.

Note that the operation of indexing the vertices in $\rho^{-1}(\upsilon)$ has an ambiguity, i.e. the choice of $\tilde{\upsilon}_1 \in \rho^{-1}(\upsilon)$ is arbitrary for each υ. We will show that this ambiguity does not affect the cohomology class of c^1.

In fact, suppose the vertices $\tilde{\upsilon}_1, \tilde{\upsilon}_2, \ldots, \tilde{\upsilon}_{W(\upsilon)}$ are re-indexed as $\tilde{\upsilon}_{1+k}, \tilde{\upsilon}_{2+k}, \ldots, \tilde{\upsilon}_{W(\upsilon)+k}$. The amount of the shift, k, depends on v, so we denote it by $k(v)$. Then, on an edge e with $\partial e = \upsilon - w$, the new 1-cochain $\overline{c}^1$ takes the following value:

$$\overline{c}^1(e) = (\alpha + k(\upsilon)) - (\beta + k(w)) = (\alpha - \beta) + (k(\upsilon) - k(w)) \in \mathbf{Z}/\gcd(W(\upsilon), W(w)).$$

Therefore, defining a 0-cochain $c^0 \in C^0(Y)$ by

$$c^0(\upsilon) = k(\upsilon),$$

we have,

$$\overline{c}^1 = c^1 + \delta(c^0).$$

Thus we have proved the following:

Lemma 6.4. *There is a well-defined cohomology class $[c^1] \in H^1_W(Y)$ associated with the pair (X, φ).*

Also the above argument has already proved the following:

Corollary 6.3. *Let $c^1 \in C^1(Y)$ be any 1-cochain representing the cohomology class $[c^1] \in H^1_W(Y)$ associated with the pair (X, φ). Then we can properly index the vertices in $\rho^{-1}(\upsilon)$ for every υ so that the associated 1-cochain is exactly the given cochain c^1.*

Suppose we have indexed the vertices of $\rho^{-1}(\upsilon)$ for every υ. In the second assertion (ii) of the following lemma, $c^1 \in C^1(Y)$ is the associated 1-cochain with this indexing. Let e be an oriented edge of Y and assume $\partial e = \upsilon - w$.

Lemma 6.5. *(i) The weight $W(e)$ is a common multiple of the weights $W(v)$, $W(w)$.*

(ii) A vertex $\tilde{v}_\alpha \in \rho^{-1}(v)$ is joined to a vertex $\tilde{w}_\beta \in \rho^{-1}(w)$ by an edge over e if and only if $\alpha - \beta = c^1(e) \in \mathbf{Z}/\gcd(W(v), W(w))$. Moreover, if $\tilde{v}_\alpha$ and $\tilde{w}_\beta$ are joined by an edge over e there are exactly $W(e)/\mathrm{lcm}(W(v), W(w))$ edges over e connecting $\tilde{v}_\alpha$ and $\tilde{w}_\beta$.

Proof. (i) is left to the reader. We will prove the first claim of (ii). The "only if" part is already proved. We will prove the converse, i.e. if $\alpha - \beta = c^1(e) \in \mathbf{Z}/\gcd$

$(W(v), W(w))$, then $\tilde{v}_\alpha$ and $\tilde{w}_\beta$ are joined by an edge over e. *Proof.* There is certainly an edge $\tilde{e}$ over e which joins $\tilde{v}_\alpha$ to a vertex over w, say $\tilde{w}_\gamma$. By the "only if" part, $\alpha - \gamma = c^1(e) \in \mathbf{Z}/\gcd(W(v), W(w))$. Thus

$$\beta \equiv \gamma \pmod{\gcd(W(v), W(w))},$$

in other words,

$$\beta = \gamma + nG, \quad G = \gcd(W(v), W(w)).$$

There are integers k, l, such that

$$kW(v) + lW(w) = G.$$

Then $\varphi^{nkW(v)}$ sends $\tilde{v}_\alpha$ to $v_{\alpha+nkW(v)} = \tilde{v}_\alpha$ (because $\alpha \in \mathbf{Z}/W(v)$) and sends $\tilde{w}_\gamma$ to

$$\tilde{w}_{\gamma+nkW(v)} = \tilde{w}_{\gamma+nG-nlW(w)} = \tilde{w}_\beta$$

because $\beta, \gamma \in \mathbf{Z}/W(w)$ and $\beta = \gamma + nG$. Therefore, $\varphi^{nkW(v)}(\tilde{e})$ connects $\tilde{v}_\alpha$ and $\tilde{w}_\beta$.

The second claim of (ii) is left to the reader. □

Let W_0 be the gcd of the weights of all the vertices of Y, and let $H_1(Y)$ be the ordinary homology group of Y. A cohomology class $c \in H^1_W(Y)$ defines a homomorphism

$$c_* : H_1(Y) \to \mathbf{Z}/W_0,$$

which sends a 1-cycle $\sum m_i e_i$ to $\sum m_i c(e_i) \in \mathbf{Z}/W_0$.

Lemma 6.6. *X is connected if and only if Y is connected and $c_* : H_1(Y) \to \mathbf{Z}/W_0$ is onto, where $c \in H^1_W(Y)$ is the cohomology class associated with (X, φ).*

The proof is left to the reader as an exercise.

The (equivariant) isomorphism class of (X, φ) is classified as follows:

Theorem 6.2. *Let (X_1, φ_1) and (X_2, φ_2) be pairs, each consisting of a graph and a (periodic) isomorphism on it. Let $Y_1 = X_1/\varphi_1$ and $Y_2 = X_2/\varphi_2$ be the respective quotient graphs. Then there exists an isomorphism*

$$\Phi : X_1 \to X_2$$

such that $\varphi_1 = \Phi^{-1}\varphi_2\Phi$ if and only if there exists a weighted isomorphism

$$\Psi : Y_1 \to Y_2$$

such that $\Psi^*(c_2) = c_1$, where

$$\Psi^* : H^1_W(Y_2) \to H^1_W(Y_1)$$

is the induced isomorphism and $c_i \in H^1_W(Y_i)$ *is the cohomology class associated with* (X_i, φ_i), $i = 1, 2$.

Proof. The "only if" part is trivial. We will prove the "if" part. We may assume that the vertices of X_1 and X_2 are properly indexed so that the 1-cochain c^1_2 associated with (X_2, φ_2) is pulled back by Ψ exactly to the 1-cochain c^1_1 associated with (X_1, φ_1), (Corollary 6.3).

Let $X_1^{(0)}$ (resp. $X_2^{(0)}$) denote the set of the vertices of X_1 (resp. X_2).

For each vertex v of Y_1, define a bijective map

$$\Phi_v : \rho_1^{-1}(v) \to \rho_2^{-1}(\Psi(v))$$

by sending a vertex $\tilde{v}_\alpha$ in $\rho_1^{-1}(v)$ to the vertex $\tilde{v}'_\alpha$ in $\rho_2^{-1}(\Psi(v))$ having the same index α as $\tilde{v}_\alpha$. This is possible because $\rho_1^{-1}(v)$ and $\rho_2^{-1}(\Psi(v))$ contain the same number of vertices (for $\Psi : Y_1 \to Y_2$ preserves the weights). The set $\{\Phi_v\}_v$ gives a bijection

$$\Phi^{(0)} : X_1^{(0)} \to X_2^{(0)}.$$

For each edge e of Y_1, there exists an edge $\tilde{e}$ of X_1 which lies over e. Let us fix $\tilde{e}$ for a while. Let $\partial e = v - w$. Suppose $\partial \tilde{e} = \tilde{v}_\alpha - \tilde{w}_\beta$, where $\tilde{v}_\alpha \in \rho_1^{-1}(v)$ and $\tilde{w}_\beta \in \rho_1^{-1}(w)$. Then there exists an edge e' of X_2, lying over $\Psi(e)$ and connecting $\tilde{v}'_\alpha$ and $\tilde{w}'_\beta$, because $\beta - \alpha = c^1_1(e) = c^1_2(\Psi(e))$, (Lemma 6.5).

Define

$$\Phi_0 : \rho_1^{-1}(e) \to \rho_2^{-1}(\Psi(e))$$

by sending $\tilde{e}$ to $\tilde{e}'$, and sending $\varphi_1^k(\tilde{e})$ to $\varphi_2^k(\tilde{e}')$ for each $k \geq 1$. Since the weights of e and $\Psi(e)$ are the same, $\varphi_1^k(\tilde{e}) = \tilde{e}$ if and only if $\varphi_2^k(\tilde{e}') = \tilde{e}'$. This means that

$$\Phi_0 : \rho_1^{-1}(e) \to \rho_2^{-1}(\Psi(e))$$

is well-defined and that it is equivariant with respect to φ_1 and φ_2.

Now the set $\{\Phi_v\}_v \cup \{\Phi_0\}_0$ gives the desired isomorphism

$$\Phi : X_1 \to X_2.$$

This completes the proof of Theorem 6.2. □

Remark 6.5. The isomorphism $\Phi : X_1 \to X_2$ obtained above makes the diagram commute:

$$\begin{array}{ccc} X_1 & \xrightarrow{\Phi} & X_2 \\ \rho_1 \downarrow & & \downarrow \rho_2 \\ Y_1 & \xrightarrow[\Psi]{} & Y_2 \end{array}$$

Combining Theorems 6.1 and 6.2, we obtain the following theorem, which is the main theorem of Part I.

Theorem 6.3. *Let*

$$f_1 : \Sigma_g^{(1)} \to \Sigma_g^{(1)} \quad \textit{and} \quad f_2 : \Sigma_g^{(2)} \to \Sigma_g^{(2)}$$

be pseudo-periodic maps of negative twist, both in superstandard form. Let

$$\pi_1 : \Sigma_g^{(1)} \to S[f_1] \quad \textit{and} \quad \pi_2 : \Sigma_g^{(2)} \to S[f_2]$$

be the respective minimal quotients.

Then there exists a homeomorphism

$$h : \Sigma_g^{(1)} \to \Sigma_g^{(2)}$$

such that $f_1 = h^{-1} f_2 h$ *if and only if there exist a numerical homeomorphism*

$$H : S[f_1] \to S[f_2]$$

and a weighted isomorphism

$$\Psi : Y[f_1] \to Y[f_2]$$

such that

(i) the following diagram commutes:

$$\begin{array}{ccc} S[f_1] & \xrightarrow{H} & S[f_2] \\ {\scriptstyle \eta_1}\downarrow & & \downarrow{\scriptstyle \eta_2} \\ Y[f_1] & \xrightarrow[\Psi]{} & Y[f_2] \end{array}$$

(ii) $\Psi^*(c[f_2]) = c[f_1]$,
where

$$\Psi^* : H^1_W(Y[f_2]) \to H^1_W(Y[f_1])$$

is the induced isomorphism, and $c[f_i] \in H^1_W(Y[f_i])$ *is the cohomology class associated with* $(\overline{X}[f_i], \varphi_{[f_i]})$, $i = 1, 2$.

Remark 6.6. Theorem 6.3 says that the triple $(S[f], Y[f], c[f])$ is a complete set of conjugacy invariants of the mapping class $[f]$ of a pseudo-periodic map $f : \Sigma_g \to \Sigma_g$ with negative twist. Triples (S, Y, c) coming from a pseudo-periodic map f of negative twist will be characterized in Part II. (See Theorem 9.2).

Part II
The Topology of Degeneration of Riemann Surfaces

In Part II, we will apply the results of Part I to study the topology of degeneration of Riemann surfaces. The main theorem of Part II is Theorem 7.2.

Chapter 7
Topological Monodromy

Definition 7.1 (Cf. [48]). A triple (M, D, ψ) is called a *degenerating family of Riemann surfaces of genus g* (abbreviated as *degenerating family of genus g*) if

(i) M is a complex surface,
(ii) $D = \{\xi \in \mathbf{C} \mid |\xi| < 1\}$,
(iii) $\psi : M \to D$ is a surjective proper holomorphic map,
(iv) for each $\xi \in D$, the fiber $F_\xi = \psi^{-1}(\xi)$ is connected, and
(v) $\psi|_{M^*} : M^* \to D^*$ is a smooth (i.e. C^∞) fiber bundle whose fiber is a Riemann surface of genus g, where $D^* = D - \{0\}$ and $M^* = M - \psi^{-1}(0)$.
If futhermore (M, D, ψ) satisfies the following condition (vi), (M, D, ψ) is said to be *minimal*.
(vi) No fiber F_ξ contains a smoothly embedded sphere of self-intersection number -1.
A fiber F_ξ, $\xi \neq 0$, will be called a *general fiber*. The central fiber F_0 may contain a singular point of ψ, in which case F_0 will be called *singular fiber*.

Let us choose a base point $\xi_0 \in D^*$, and let $l : [0, 2\pi] \to D^*$ be a loop based at ξ_0 ($l(0) = l(2\pi) = \xi_0$) which turns around 0 once in the positive direction. Then there is a continuous family of (orientation-preserving) homeomorphisms

$$h_\theta : \Sigma_g \to F_{l(\theta)} \quad (0 \leq \theta \leq 2\pi),$$

where Σ_g is a fixed surface of genus g.

Definition 7.2. $f = h_0^{-1} h_{2\pi} : \Sigma_g \to \Sigma_g$ is called a *monodromy homeomorphism* of the family (M, D, ψ).

Given a degenerating family (M, D, ψ) of genus g, a monodromy homeomorphism

$$f : \Sigma_g \to \Sigma_g$$

Y. Matsumoto and J.M. Montesinos-Amilibia, *Pseudo-periodic Maps and Degeneration of Riemann Surfaces*, Lecture Notes in Mathematics 2030,
DOI 10.1007/978-3-642-22534-5_7,

is well-defined up to *isotopy* and *conjugation*. More precisely, let $\widehat{\mathcal{M}_g}$ denote the set of conjugacy classes in $\mathcal{M}_g$ (= the mapping class group of Σ_g). Then the conjugacy class $\langle f \rangle \in \widehat{\mathcal{M}_g}$ to which the mapping class $[f]$ of f belongs is determined by (M, D, ψ) and is independent of the various choices involved. See [45].

Definition 7.3. $\langle f \rangle \in \widehat{\mathcal{M}_g}$ is called the *topological monodromy* of (M, D, ψ).

Henceforth we will assume $g \geq 2$.

The main effort of Chap. 7 will be devoted to the proof of

Theorem 7.1 ([19, 26, 27, 58]). *A monodromy homeomorphism*

$$f : \Sigma_g \to \Sigma_g$$

of a degenerating family (M, D, ψ) of genus g is a pseudo-periodic map of negative twist. (See Chap. 3 for the definition of a pseudo-periodic map of negative twist.)

This is proved by Earle and Sipe [19, Sect. 7] using Teichmüller space theory. See [26, 58] for preceding important results. We give our topological proof here because it will clarify the topological structure of (M, D, ψ) whose understanding is a key step to the main result of Part II.

Definition 7.4. Two degenerating families of genus g, (M, D, ψ) and (M', D', ψ'), are *topologically equivalent* (or *have the same topological type*) if there exist orintation-preserving homeomorphisms $\Lambda : M \to M'$ and $\lambda : D \to D'$ such that

(i) $\lambda(0) = 0$, and
(ii) $\psi' \Lambda = \lambda \psi$.

Let $\mathcal{S}_g$ denote the set of all topological types of degenerating families of genus g which are *minimal*. Then since topologically equivalent degenerating families have the same topological monodromy, the map $\rho : \mathcal{S}_g \to \widehat{\mathcal{M}_g}$ sending (M, D, ψ) to its topological monodromy $\langle f \rangle$ is well-defined. By Theorem 7.1, the image of ρ is contained in the set of $\mathcal{P}_g^-$ of conjugacy calsses represented by pseudo-periodic maps of negative twist.

The main result of Part II is the following.

Theorem 7.2. *The map $\rho : \mathcal{S}_g \to \mathcal{P}_g^-$ is bijective for $g \geq 2$.*

The proof will be given in Chap. 9. By Dehn and Nielsen [53, Sects. 22, 23], $\mathcal{M}_g \cong Aut(\pi_1 \Sigma_g)/Inn(\pi_1 \Sigma_g)$. Thus we have

Corollary 7.1. *If (M, D, ψ) is minimal, the action of the monodromy on the fundamental group $\pi_1 \Sigma_g$ modulo inner automorphisms determines the topological type of (M, D, ψ).*

It has been pointed out by Namikawa and Ueno [49] that the action of the monodromy on the homology group $H_1(\Sigma_g; \mathbf{Z})$ does not necessarily determine the topological type of (M, D, ψ). The algebraic and combinatorial description of the action of the monodromy on $\pi_1 \Sigma_g$ has been given in the case of stable curves by Asada, Matsumoto and Oda [4], for number theoretic purposes.

As a special case of Corollary 7.1, we have a topological proof of

Corollary 7.2 (Cf. [26, 48]). *If* (M, D, ψ) *is minimal and if the action of monodromy on* $\pi_1 \Sigma_g$ *is an inner automorphism, then* F_0 *is non-singular.*

Strictly speaking, the implication of Corollary 7.1 is this: under the condition of Corollary 7.2, F_0 is *topologically* equivalent to a non-sigular fiber. However, this actually assures that F_0 is *analytically* non-singular. See Corollary 8.4.

Although the authors could not find any references stating the result clearly, Corollary 7.2 might not be a new result because it can be immediately proved by combining Theorem 1.6 of [26] and Lemma 3.2 of [48].

Corollary 7.3. *Given a pseudo-periodic map of negative twist* $f : \Sigma_g \to \Sigma_g$*, there exists a degenerating family* (M, D, ψ) *whose monodromy homeomorphism around* F_0 *is equal (up to isotopy and conjugation) to* f*.*

Essentially the same result as Corollary 7.3 is announced by Earle and Sipe [19, p. 92].

7.1 Proof of Theorem 7.1

In what follows, $F_0 = \psi^{-1}(0)$ will have only *normal crossings* (nodes). This can be attained by blowing up F_0 (without affecting the topological monodromy around it), see [33].

The closure of a connected component of $F_0 - \{nodes\}$ is called an *irreducible component* of F_0. Each irreducible component Θ is a smoothly immersed closed Riemann surface, and F_0 is the union of these irreducible components:

$$F_0 = \Theta_1 \cup \Theta_2 \cup \cdots \cup \Theta_s.$$

Let p be a generic point of Θ_i (namely, a point which is not a node). Then we can find local coordinates (z_1, z_2) of M such that $(z_1(p), z_2(p)) = (0, 0)$ and $\psi : M \to D$ is locally written as

$$\psi(z_1, z_2) = (z_2)^{m_i}.$$

The positive integer m_i does not depend on the choice of p and is determined by Θ_i. The number of m_i is called the *multiplicity* of Θ_i, and the coordinates (z_1, z_2) satisfying the above condition are called *admissible coordinates* centered at p. To make the multiplicities explicit, it will be convenient to denote F_0 as a divisor:

$$F_0 = m_1\Theta_1 + m_2\Theta_2 + \cdots + m_s\Theta_s.$$

Let p be a crossing point of Θ_i and Θ_j (the case $\Theta_i = \Theta_j$ is not excluded), then we can find local coordinates (z_1, z_2) of M such that $(z_1(p), z_2(p)) = (0, 0)$ and $\psi : M \to D$ is locally written as

$$\psi(z_1, z_2) = (z_1)^{m_j} (z_2)^{m_i}.$$

Again such coordinates (z_1, z_2) will be called *admissible coordinates* centered at p.

The plan of the proof is as follows. We decompose a general fiber F_ξ which is near to F_0 into two parts A_ξ and B_ξ. A_ξ is a union of annuli, and B_ξ is a union of compact surfaces. A monodromy homeomorphism $f : F_\xi \to F_\xi$ will be constructed so that it preserves this decomposition. f acts on B_ξ as a periodic map, while it gives twisting to A_ξ. This will prove that f is a pseudo-periodic map. The proof of the negativity of twisting requires a closer look, which will be taken in the second half of Chap. 7.

At each node $p \in F_0$, we will construct a polydisk Δ_p. Let Δ denote the union $\cup_p \Delta_p$. Then $A_\xi = F_\xi \cap \Delta$, and $B_\xi = F_\xi \cap (M - Int(\Delta))$. We will also construct a system of tubular neighborhoods $\{N_i\}_{i=1}^s$ of the irreducible components $\{\Theta_i\}_{i=1}^s$. This is to control the behaviour of f outside Δ.

7.2 Construction of Δ and $\{N_i\}_{i=1}^s$

Let p be a node of F_0. We fix admissible coordinates (z_1, z_2) centered at p. Define a polydisk Δ_p by

$$\Delta_p = \{(z_1, z_2) \mid |z_1| \leq \varepsilon,\ |z_2| \leq \varepsilon\},$$

where ε is a sufficiently small number (> 0) chosen independently of p. Let Δ be the union $\cup_p \Delta_p$, p running over all the nodes of F_0.

Let N_i be a tubular neighborhood of Θ_i, $i = 1, 2, \ldots, s$. N_i has the structure of a D^2-bundle over Θ_i. We choose the projections $\pi_i : N_i \to \Theta_i$ so as to be "compatible" with the fixed admissible coordinates in Δ_p, in other words, if p is an intersection point of Θ_i and Θ_j, the projections π_i and π_j should satisfy

$$\pi_i(z_1, z_2) = z_1, \quad \pi_j(z_1, z_2) = z_2.$$

Also N_i must be very thin. To fix the idea, we will assume that the radius of the fiber is $\leq \varepsilon/2$. (See Fig. 7.1)

Finally we make admissible coordinates outside Δ *compatible* with the projections, π_i's, as follows. Let p be a generic point of Θ_i, (z_1, z_2) admissible coordinates centered at p. We deform (z_1, z_2) by a C^∞-isotopy which moves them in the z_1-direction so that with respect to the resulting coordinates the projection $\pi_i : N_i \to \Theta_i$ is locally written as

$$\pi_1(z_1, z_2) = z_1,$$

and the equation $\psi(z_1, z_2) = (z_2)^{m_i}$ be still valid.

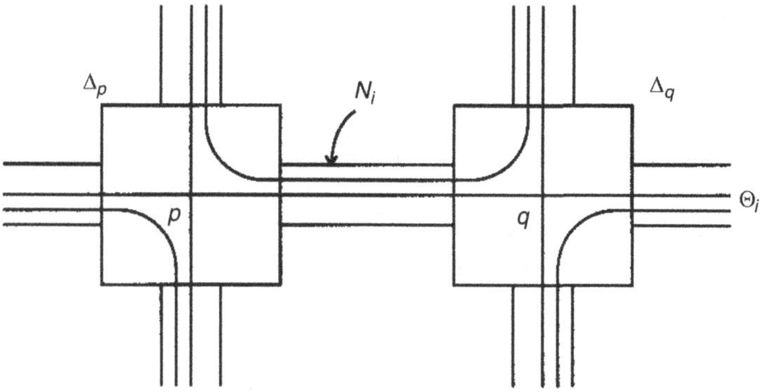

Fig. 7.1 Polydisks Δ_p, Δ_q, and a tubular neighborhood N_i of Θ_i

The new coordinates (z_1, z_2) will no longer give the complex structure of M, but our argument in Chap. 7 will be in the topological category, so this will not cause any essential difficulty.

7.3 The Decomposition $F_\xi = A_\xi \cup B_\xi$

If $\xi \neq 0$ is a complex number sufficiently close to 0, we define

$$A_\xi = F_\xi \cap \Delta, \quad B_\xi = F_\xi \cap (M - Int(\Delta)).$$

7.4 Construction of a Monodromy Homeomorphism

Let $e(\theta)$ denote $\exp(\sqrt{-1}\theta)$. Let δ be a small positive real number. We will construct a continuous family of homeomorphisms

$$h_\theta : F_\delta \to F_{e(\theta)\delta} \quad (0 \leq \theta \leq 2\pi)$$

starting with $h_0 = id$. Then $f = h_{2\pi} : F_\delta \to F_\delta$ will be a monodromy homeomorphism of (M, D, ψ).

The construction will be done separately on the A_δ-part and on the B_δ-part. We will begin with B_δ-part.

For a general (small) ξ, let $B_{i,\xi}$ denote $F_\xi \cap (N_i - Int(\Delta))$. Note that $B_\xi = \bigcup_{i=1}^s B_{i,\xi}$ (disjoint union). Let $\check{\Theta}_i = \Theta_i - (\Theta_i \cap Int(\Delta))$. $\check{\Theta}_i$ is a compact surface, generally with boundary. Then $B_{i,\xi}$ is an m_i-fold cyclic covering over $\check{\Theta}_i$ with the projection $\pi_i|_{B_{i,\xi}} : B_{i,\xi} \to \check{\Theta}_i$ (m_i = the multiplicity of Θ_i) because, around each point $p \in \check{\Theta}_i$, $B_{i,\xi}$ and $\check{\Theta}_i$ are given by

$$(z_2)^{m_i} = \xi \quad \text{and} \quad z_2 = 0$$

respectively (with respect to admissible coordinates) and the projection π_i is given by $\pi_i(z_1, z_2) = z_1$.

Now consider the special value $\xi = e(\theta)\delta$.

For each $\theta \in [0, 2\pi]$, define $h_\theta|_{B_{i,\delta}} : B_{i,\delta} \to B_{i,e(\theta)\delta}$ locally by

$$h_\theta(z_1, z_2) = \left(z_1, e\left(\frac{\theta}{m_i}\right) z_2\right)$$

using admissible coordinates. If $(z_2)^{m_i} = \delta$, then

$$[e(\theta/m_i)\, z_2]^{m_i} = e(\theta)\delta.$$

Thus h_θ sends $B_{i,\delta}$ into $B_{i,e(\theta)\delta}$.

Note that $h_\theta|_{B_{i,\delta}}$ respects the 2-disk fibers of $\pi_i : N_i \to \check{\Theta}_i$ because (z_1, z_2) are compatible with π_i. Also note that the second coordinates z_2, z_2', in two overlapping systems of admissible coordinates (z_1, z_2) and (z_1', z_2'), differ by an m_i-th root of unity: $z_2 = e(2\pi k/m_i)z_2'$ (for some $k \in \mathbf{Z}$), because $z_2^{m_i} = \psi(z_1, z_2) = \psi(z_1', z_2') = (z_2')^{m_i}$. Thus the local definition of $h_\theta|_{B_{i,\delta}}$ is independent of the choice of admissible coodinates, and $h_\theta|B_{i,\delta}$ is globally defined over $B_{i,\delta}$.

$h_\theta|_{B_{i,\delta}} : B_{i,\delta} \to B_{i,e(\theta)\delta}$ is a homeomorphism which depends continuously on θ.

Clearly $h_0|_{B_{i,\delta}} = id$, and $h_{2\pi}|_{B_{i,\delta}} : B_{i,\delta} \to B_{i,\delta}$ is a periodic map of order m_i, which is a covering translation of $B_{i,\delta} \to \check{\Theta}_i$. We define a continuous family of homeomorphisms

$$h_\theta|_{B_\delta} : B_\delta \to B_{e(\theta)\delta}$$

to be the disjoint union $\cup_{i=1}^{s}(h_\theta|_{B_{i,\delta}})$. Then $h_0|_{B_\delta} = id$, and $h_{2\pi}|_{B_\delta}$ is a periodic map of order $\mathrm{lcm}(m_1, m_2, \ldots, m_s)$.

The construction on the B_δ-part is completed.

For a general (small) ξ, let $A_{p,\xi}$ denote $F_\xi \cap \Delta_p$. If p is an intersection point of Θ_i and Θ_j, then (with admissible coordinates)

$$A_{p,\xi} = \{(z_1, z_2) \mid (z_1)^{m_j}(z_2)^{m_i} = \xi,\ |z_1| \le \varepsilon,\ |z_2| \le \varepsilon\},$$

and $A_\xi = \cup_p A_{p,\xi}$ (disjoint union). *Each $A_{p,\xi}$ is a disjoint union of annuli.* To see this, consider an annulus on $\Delta_p \cap \Theta_i$, defined by

$$\{(z_1, 0) \mid \varepsilon' \le |z_1| \le \varepsilon\},$$

where $\varepsilon' = (|\xi|/\varepsilon^{m_i})^{1/m_j}$. See Fig. 7.2.

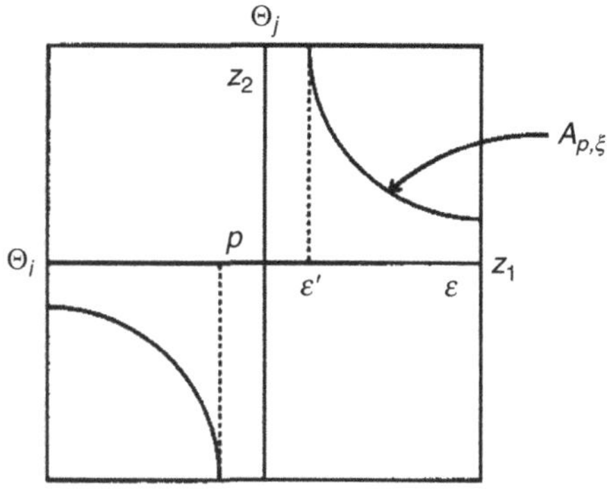

Fig. 7.2 $A_{p,\xi}$ is a disjoint union of annuli

$A_{p,\xi}$ is an m_i-fold cyclic covering over this annulus, with the projection $(z_1, z_2) \longmapsto (z_1, 0)$. Thus $A_{p,\xi}$ is a disjoint union of annuli.

The boundary of $A_{p,\xi}$ is divided into two parts:

$$A_{p,\xi} \cap \{(z_1, z_2) \mid |z_1| = \varepsilon\}$$

and

$$A_{p,\xi} \cap \{(z_1, z_2) \mid |z_2| = \varepsilon\},$$

which we call the *i-boundary* and the *j-boundary*, respectively.

The i-boundary is a torus link of type (m_i, m_j) in $\partial\Delta_p \approx S^3$ as well as the j-boundary. The number of the connected components of such a link is $\gcd(m_i, m_j)$, which is denoted by m.

$A_{p,\xi}$ is homeomorphic to (i-boundary) $\times [\varepsilon', \varepsilon]$. Thus $A_{p,\xi}$ consists of m annuli.

Now putting $\xi = \delta$ (sufficiently small, for instance $\leq 1/(2\varepsilon)$), we will define a homeomorphism

$$h_\theta|_{A_{p,\delta}} : A_{p,\delta} \to A_{p,e(\theta)\delta}.$$

Let $t = t(|z_1|, |z_2|)$ be an auxiliary, real-valued function on Δ'_p $(= \Delta_p - \{(z_1, z_2) \mid |z_1| = |z_2| = \varepsilon\})$ defined by

$$t = \begin{cases} \frac{\varepsilon - |z_2|}{2(\varepsilon - |z_1|)} & \text{if} \quad |z_1| \leq |z_2| \leq \varepsilon, \\ 1 - \frac{\varepsilon - |z_1|}{2(\varepsilon - |z_2|)} & \text{if} \quad \varepsilon \geq |z_1| \geq |z_2|. \end{cases}$$

(See Fig. 7.3)

Then the homeomorphism $h_\theta|_{A_{p,\delta}}$ is defined as follows:

$$h_\theta(z_1, z_2) = \left(e\left(\frac{(1-t)\theta}{m_j}\right) z_1, \, e\left(\frac{t\theta}{m_i}\right) z_2 \right).$$

It is easy to see that $h_\theta|_{A_{p,\delta}}$ is in fact a homeomorphism of $A_{p,\delta}$ onto $A_{p,e(\theta)\delta}$, and that it preserves the value of the function $t(|z_1|, |z_2|)$. Let $C_{p,e(\theta)\delta}$ denote the set of

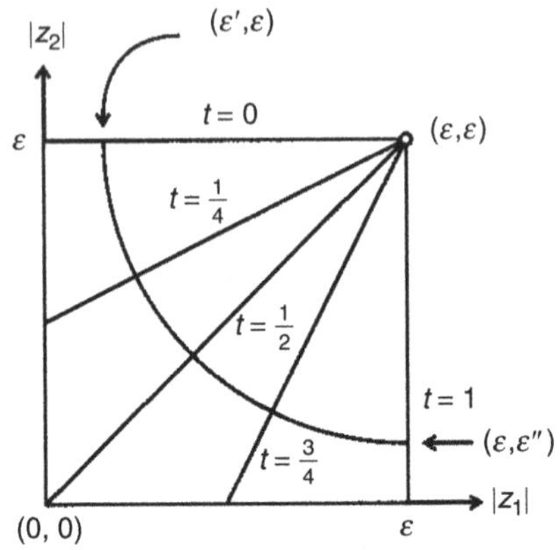

Fig. 7.3 $\varepsilon' = (\delta/\varepsilon^{m_i})^{1/m_j}$; $\varepsilon'' = (\delta/\varepsilon^{m_j})^{1/m_i}$

the center-lines of the annuli $A_{p,e(\theta)\delta}$:

$$C_{p,e(\theta)\delta} = A_{p,e(\theta)\delta} \bigcap \{(z_1, z_2) \mid |z_1| = |z_2|\}$$
$$= A_{p,e(\theta)\delta} \bigcap \left\{(z_1, z_2) \mid t = \frac{1}{2}\right\}.$$

Then $h_\theta|_{A_{p,\delta}}$ sends $C_{p,\delta}$ onto $C_{p,e(\theta)\delta}$ homeomorphically. Cleary, $h_\theta|_{A_{p,\delta}}$ depends continuously on θ, and $h_0|_{A_{p,\delta}} = id$.

On the i-boundary of $A_{p,\delta}$, where $t = 1$, we have

$$h_\theta(z_1, z_2) = \left(z_1, e\left(\frac{\theta}{m_i}\right) z_2\right),$$

which coincides with $h_\theta|_{B_{i,\delta}}$ on $A_{p,\delta} \cap B_{i,\delta}$. On the j-boundary, where $t = 0$, we have

$$h_\theta(z_1, z_2) = \left(e\left(\frac{\theta}{m_j}\right) z_1, z_2\right),$$

which coincides with $h_\theta|_{B_{j,\delta}}$ on $A_{p,\delta} \cap B_{j,\delta}$. Since $A_\delta = \cup_p A_{p,\delta}$, we define a continuous family of homeomorphisms

$$h_\theta|_{A_\delta} : A_\delta \to A_{e(\theta)\delta}$$

as the disjoint union $\bigcup_p (h_\theta|_{A_{p,\delta}})$.

$h_\theta|_{A_\delta}$ and $h_\theta|_{B_\delta}$ coincide on their common boundaries, thus they give a continuous family of homeomorphisms

$$h_\theta = h_\theta|_{A_\delta} \bigcup h_\theta|_{B_\delta} : F_\delta \to F_{e(\theta)\delta} \quad (0 \le \theta \le 2\pi),$$

with $h_0 = id$. The monodromy homeomorphism $f : F_\delta \to F_\delta$ is the final stage of $h_\theta : f = h_{2\pi}$.

Let $\mathscr{C}_\delta = \bigcup_p C_{p,\delta}$ be the disjoint union of the center-lines of the annuli. Then f preserves $\mathscr{C}_\delta$, and

$$f|_{F_\delta - \mathscr{C}_\delta} : F_\delta - \mathscr{C}_\delta \to F_\delta - \mathscr{C}_\delta$$

is isotopic to a periodic map because $F_\delta - \mathscr{C}_\delta$ is the union of B_δ and open collars $A_\delta - \mathscr{C}_\delta$ on ∂B_δ and the restriction $f|_{B_\delta} = f_{2\pi}|_{B_\delta}$ is a periodic map as we saw before.

We have proved the following:

Lemma 7.1. *The monodromy homeomorphism $f : F_\delta \to F_\delta$ is a pseudo-periodic map.*

7.5 Negativity of Screw Numbers

Fix an intersection point p of Θ_i and Θ_j. We look at the action of f on $A_{p,\delta}$ more closely.

Define positive reals ε', ε'' by

$$\varepsilon' = \left(\frac{\delta}{\varepsilon^{m_i}}\right)^{1/m_j}, \quad \varepsilon'' = \left(\frac{\delta}{\varepsilon^{m_j}}\right)^{1/m_i}.$$

Then $(\varepsilon, \varepsilon'') \in \Delta_p$ is a point of the i-boundary of $A_{p,\delta}$. Let $A^0_{p,\delta}$ denote the component of $A_{p,\delta}$ which contains the point $(\varepsilon, \varepsilon'')$.

We will construct a parametrization

$$\phi : [0, 1] \times S^1 \to A^0_{p,\delta}.$$

For this, note that for each $t \in [0, 1]$, there is a unique pair of reals (r, s) satisfying

$$0 \le r < \varepsilon, \quad 0 \le s < \varepsilon, \quad r^{m_j} s^{m_i} = \delta, \quad \text{and} \quad t(r, s) = t,$$

where $t(|z_1|, |z_2|)$ is the function introduced just before the definition of $h_\theta|_{A_{p,\delta}}$. In fact, the point (r, s) is the intersection of the curve $r^{m_j} s^{m_i} = \delta$ and the line $t(r, s) = t$. (See Fig. 7.3.) Since (r, s) is uniquely determined by $t \in [0, 1]$, we denote it by $(r(t), s(t))$. Then

$$(r(0), s(0)) = (\varepsilon', \varepsilon), \quad (r(1), s(1)) = (\varepsilon, \varepsilon''),$$

and $(r(t), s(t))$ determines the curve $[0, 1] \to A^0_{p,\delta}$ joining $(\varepsilon', \varepsilon) \in (j - \textit{boundary})$ to $(\varepsilon, \varepsilon'') \in (i - \textit{boundary})$.

Let n_i, n_j be the positive integers given by

$$n_i = \frac{m_i}{m}, \quad n_j = \frac{m_j}{m},$$

where $m = \gcd(m_i, m_j)$.

We define the parametrization

$$\phi : [0,1] \times S^1 \to A^0_{p,\delta}$$

as follows:

$$\phi(t, x) = (e(2\pi n_i x) r(t),\ e(-2\pi n_j x) s(t)),$$

where $t \in [0,1]$ and $x \in S^1 = \mathbf{R}/\mathbf{Z}$. If $x = 0$, $\phi(t, 0)$ $(0 \leq t \leq 1)$ is the curve $(r(t), s(t))$, and if $t = 1$, $\phi(1, x)$ $(0 \leq x \leq 1)$ is a torus knot of type (n_i, n_j), which is the i-boundary of $A^0_{p,\delta}$. It is readily seen that $\phi : [0,1] \times S^1 \to A^0_{p,\delta}$ is a homeomorphism.

For an integer a, let us denote by $A^a_{p,\delta}$ the component of $A_{p,\delta}$ containing the point $(\varepsilon, e(2\pi/m_i)^a \varepsilon'')$. Then, by the property of a torus link of type (m_i, m_j),

$$A^a_{p,\delta} = A^b_{p,\delta} \quad \text{iff} \quad a \equiv b \pmod{m}.$$

The monodromy homeomorphism

$$f|_{A_{p,\delta}} (= h_{2\pi}|_{A_{p,\delta}}) : A_{p,\delta} \to A_{p,\delta}$$

satisfies

$$(f|_{A_{p\delta}}) \left(\varepsilon,\ e\left(\frac{2\pi}{m_i}\right)^a \varepsilon'' \right) = \left(\varepsilon,\ e\left(\frac{2\pi}{m_i}\right)^{a+1} \varepsilon'' \right),$$

so $f|_{A_{p,\delta}}$ cyclically permutes the m components of $A_{p,\delta}$:

$$A^0_{p,\delta},\ A^1_{p,\delta},\ \ldots,\ A^{m-1}_{p,\delta}.$$

The m-th power $(f|_{A_{p,\delta}})^m$ maps each component to itself. Using the parametrization

$$\phi : [0,1] \times S^1 \to A^0_{p,\delta},$$

$$(f|_{A_{p,\delta}})^m : A^0_{p,\delta} \to A^0_{p,\delta}$$

is described as follows:

$$(f|_{A_{p,\delta}})^m \phi(t, x) = \left(e\left(\frac{2\pi(1-t)}{m_j}\right)^m e(2\pi n_i x) r(t),\right.$$
$$\left. \times\ e\left(\frac{2\pi t}{m_i}\right)^m e(-2\pi n_j x) s(t) \right).$$

To simplify this expression, we introduce positive integers k, l by

$$n_j k \equiv 1 \pmod{n_i}, \quad 0 < k \leq n_i, \quad \text{and}$$
$$n_i l \equiv 1 \pmod{n_j}, \quad 0 < l \leq n_j.$$

(Note that $k = n_i$ if and only if $n_i = 1$. Similarly for l.) Then

$$e\left(\frac{2\pi(1-t)}{m_j}\right)^m e(2\pi n_i x) = e\left(\frac{2\pi(1-t)}{n_j} + 2\pi n_i x\right)$$
$$= e\left(2\pi n_i \left(x + \frac{l}{n_j} - \frac{t}{n_i n_j}\right)\right),$$

and

$$e\left(\frac{2\pi t}{m_i}\right)^m e(-2\pi n_j x) = e\left(\frac{2\pi t}{n_i} - 2\pi n_j x\right)$$
$$= e\left(-2\pi n_j \left(x + \frac{l}{n_j} - \frac{t}{n_i n_j}\right)\right).$$

Thus we obtain the following description:

$$(f|_{A_{p,\delta}})^m \phi(t, x) = \phi\left(t, x + \frac{l}{n_j} - \frac{t}{n_i n_j}\right). \qquad (*)$$

Claim. $k/n_i + l/n_j = 1 + 1/(n_i n_j)$

This claim is nothing else than Claim (J) just after the statement of Lemma 5.2 (Chap. 5).

Put $t = 0$ in $(*)$, then

$$(f|_{A_{p,\delta}})^m \phi(0, x) = \phi\left(0, x + \frac{l}{n_j}\right). \qquad (*0)$$

Put $t = 1$ in $(*)$ and apply Claim, then

$$(f|_{A_{p,\delta}})^m \phi(1, x) = \phi\left(1, x - \frac{k}{n_i}\right). \qquad (*1)$$

We have proved the following:

Theorem 7.3 (Cf. [17, Sect. 3]). *Let p be an intersection of Θ_i and Θ_j whose multiplicities are m_i and m_j respectively. Let m denote* $\gcd(m_i, m_j)$. *Then*

(i) $A_{p,\delta} = F_\delta \cap \Delta_p$ consists of m annuli,

$$A^0_{p,\delta},\ A^1_{p,\delta},\ \ldots,\ A^{m-1}_{p,\delta},$$

and the monodromy

$$f|_{A_{p,\delta}} : A_{p,\delta} \to A_{p,\delta}$$

cyclically permutes these annuli.

Fig. 7.4 $n_i = 3; k = 1$; screw number $= -1/12$

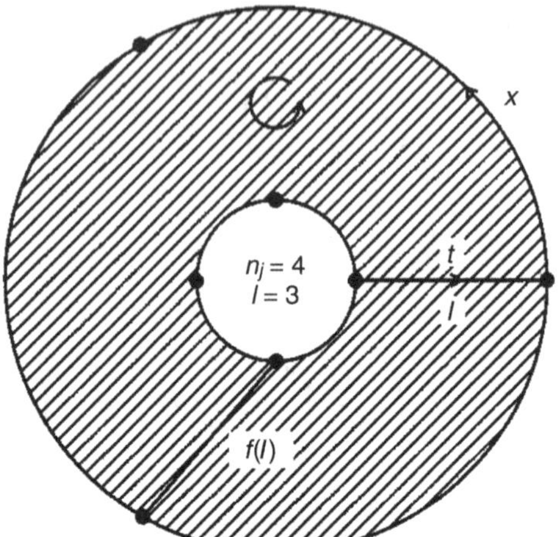

(ii) *The m-th power $(f|_{A_{p,\delta}})^m$ maps each annulus onto itself preserving the i- and j-boundaries.*
(iii) *Set $n_i = m_i/m$ and $n_j = m_j/m$. Define the integers k, l by $n_j k \equiv 1 \pmod{n_i}$, $0 < k \leq n_i$, and $n_i l \equiv 1 \pmod{n_j}$, $0 < l \leq n_j$. Then by a certain parametrization $\phi_a : [0,1] \times S^1 \to A^a_{p,\delta}$, the action of $(f|_{A_{p,\delta}})^m : A^a_{p,\delta} \to A^a_{p,\delta}$ is described as*

$$(f|_{A_{p,\delta}})^m \phi_a(t, x) = \phi_a\left(t, x + \frac{l}{n_j} - \frac{t}{n_i n_j}\right),$$

in particular on the i-boundary, as

$$x \longmapsto x - \frac{k}{n_i},$$

and on the j-boundary, as

$$x \longmapsto x + \frac{l}{n_j}.$$

(iv) *$s(A^a_{p,\delta}) = -1/(n_i n_j)$, where $s(A^a_{p,\delta})$ is the screw number of $f|_{A_{p,\delta}} : A_{p,\delta} \to A_{p,\delta}$ in $A^a_{p,\delta}$.*

Clemens [17, Sect. 3] described the monodromy at a normal crossing point in general dimensions. The above theorem is regarded as a detailed version of his result in dimension two.

By (iii) of Theorem 7.3, $f|_{A_{p,\delta}}$ is a linear twist. Thus the screw number $s(A^a_{p,\delta})$ is the coefficient of t, that is, $-1/(n_i n_j)$. See Corollary 2.2. (Beware that the direction of x in Chap. 2 is opposite to the one here. See Fig. 7.4, and remark after the example below.)
This proves (iv) of Theorem 7.3.

Example 7.1. Figure 7.4 illustrates the action of $(f|_{A_{p,\delta}})^m$ in the case $m_i = 3$, $m_j = 4$; here $m = 1$.

Remark 7.1. The orientation of $A_{p,\delta}$ in Fig. 7.4 is the natural one of the complex curve $(z_1)^{m_j}(z_2)^{m_i} = \delta$, which coincides with the orientation induced on $A_{p,\delta}$ from the natural orientation of the z_1-axis by the projection

$$A_{p,\delta} \to \{(z_1, 0) \mid \varepsilon' \le |z_1| \le \varepsilon\}, \quad (z_1, z_2) \mapsto (z_1, 0).$$

See Fig. 7.2.

7.6 Completion of the Proof of Theorem 7.1

Let us return to the decomposition $F_\delta = A_\delta \cup B_\delta$. Let $\mathscr{C}_\delta$ be the disjoint union of center-lines of A_δ. As we did in the proof of Lemma 2.1, an admissible system of cut curves is obtained from $\mathscr{C}_\delta$ by deleting inessential curves, and inserting (non-amphidrome or amphidrome) curves between parallel curves which are to be deleted. In the latter case, the screw numbers of the deleted curves (which are negative by Theorem 7.3 (iv)) are inherited by the inserted curves. Thus in the resulting admissible system, each curve has negative screw number. This completes the proof of Theorem 7.1. □

Let (M, D, ψ) be a degenerating family of genus $g \ge 2$. We assume as before that the central fiber F_0 has only normal crossings. Let $f = h_{2\pi} : F_\delta \to F_\delta$ be the monodromy homeomorphism constructed earlier in this chapter.

Theorem 7.4. *$f : F_\delta \to F_\delta$ is in superstandard form, and there exists a pinched covering $\pi : F_\delta \to F_0$ which is a generalized quotient of f (in the sense of Chap. 3).*

Proof. We continue to use the same notation as in the proof of Thorem 7.1. Recall the following decomposition of F_δ and F_0:

$$F_\delta = \bigcup_p A_{p,\delta} \cup \bigcup_i B_{i,\delta}$$

$$F_0 = \bigcup_p A_{p,0} \cup \bigcup_i \check{\Theta}_i$$

The pinched covering $\pi : F_\delta \to F_0$ will be constructed according to this decomposition, firstly on $B_{i,\delta}$ and then on $A_{p,\delta}$.

As a matter of fact, the construction $\pi|_{B_{i,\delta}} : B_{i,\delta} \to \check{\Theta}_i$ has already been done. It is simply the restriction to $B_{i,\delta}$ of the projection map $\pi_i : N_i \to \Theta_i$ of a tubular neighborhood N_i. In terms of admissible coordinates (z_1, z_2) centered at a generic point of $\check{\Theta}_i$ which are compatible with π_i, the projection $\pi|_{B_{i,\delta}}$ is given by

$$\pi(z_1, z_2) = z_1.$$

Also $B_{i,\delta}$ is locally written as $(z_2)^{m_i} = \delta$ and $\check{\Theta}_i$ as $z_2 = 0$; m_i being the multiplicity of Θ_i. Thus $\pi|_{B_{i,\delta}} : B_{i,\delta} \to \check{\Theta}_i$ is an m_i-fold cyclic covering. Next let us consider $A_{p,\delta}$. Let p be an intersection point of Θ_i and Θ_j whose multiplicities are m_i and m_j, respectively. The polydisk Δ_p was defined by

$$\Delta_p = \{(z_1, z_2) \mid |z_1| \leq \varepsilon, \ |z_2| \leq \varepsilon\},$$

where (z_1, z_2) are admissible coordinates centered at p. Also

$$A_{p,\delta} = F_\delta \cap \Delta_p = \{(z_1, z_2) \mid |z_1| \leq \varepsilon, \ |z_2| \leq \varepsilon, \ (z_1)^{m_j}(z_2)^{m_i} = \delta\},$$

and $A_{p,0}$ is a closed nodal neighborhood of p consisting of two banks $D_1 = \{(z_1, 0) \mid |z_1| \leq \varepsilon\}$ and $D_2 = \{(0, z_2) \mid |z_2| \leq \varepsilon\}$.

We will define $\pi|_{A_{p,\delta}} : A_{p,\delta} \to A_{p,0}$ as follows:

$$\pi(z_1, z_2) = \begin{cases} \left(0, \ \frac{\varepsilon(|z_2| - |z_1|)}{\varepsilon - |z_1|} \cdot \frac{z_2}{|z_2|}\right) \in D_2 & \text{if} \quad |z_1| \leq |z_2| \leq \varepsilon, \\ \left(\frac{\varepsilon(|z_1| - |z_2|)}{\varepsilon - |z_2|} \cdot \frac{z_1}{|z_1|}, \ 0\right) \in D_1 & \text{if} \quad \varepsilon \geq |z_1| \geq |z_2|. \end{cases}$$

To see the geometric nature of this map, recall the definition of the parametrization

$$\phi : [0, 1] \times S^1 \to A^0_{p,\delta},$$

introduced before Theorem 7.3:

$$\phi(t, x) = \left(e(2\pi n_i x) r(t), \ e(-2\pi n_j x) s(t)\right),$$

where $e(\theta) = \exp(\sqrt{-1}\theta)$, $t \in [0, 1]$, $x \in S^1 = \mathbf{R}/\mathbf{Z}$, $n_i = m_i/m$ and $n_j = m_j/m$, m being $\gcd(m_i, m_j)$. The pair of functions $(r(t), s(t))$ satisfies

$$r(t)^{m_j} s(t)^{m_i} = \delta \quad \text{and} \quad t(r(t), s(t)) = t,$$

where the function $t(|z_1|, |z_2|)$ was defined on $\Delta_p - \{(z_1, z_2) \mid |z_1| = |z_2| = \varepsilon\}$ by setting

$$t(|z_1|, |z_2|) = \begin{cases} \frac{\varepsilon - |z_2|}{2(\varepsilon - |z_1|)} \quad \left(\leq \frac{1}{2}\right) & \text{if} \quad |z_1| \leq |z_2| \leq \varepsilon, \\ 1 - \frac{\varepsilon - |z_1|}{2(\varepsilon - |z_2|)} \quad \left(\geq \frac{1}{2}\right) & \text{if } \varepsilon \geq |z_1| \geq |z_2|. \end{cases}$$

(See Fig. 7.3).

Then observing that

$$\frac{\varepsilon(|z_2| - |z_1|)}{\varepsilon - |z_1|} = 2\varepsilon\left(\frac{1}{2} - t(|z_1|, |z_2|)\right) \quad \text{if} \quad |z_1| \leq |z_2| \leq \varepsilon,$$

and that

$$\frac{\varepsilon(|z_1| - |z_2|)}{\varepsilon - |z_2|} = 2\varepsilon\left(t(|z_1|, |z_2|) - \frac{1}{2}\right) \quad \text{if} \quad \varepsilon \geq |z_1| \geq |z_2|,$$

we have

$$\pi\phi(t, x) = \begin{cases} \left(0, 2\varepsilon\left(\frac{1}{2} - t\right)e(-2\pi n_j x)\right), & 0 \leq t \leq \frac{1}{2}, \\ \left(2\varepsilon\left(t - \frac{1}{2}\right)e(2\pi n_i x), 0\right), & \frac{1}{2} \leq t \leq 1. \end{cases}$$

Thus the map

$$\pi|_{A_{p,\delta}} : A_{p,\delta} \to A_{p,0}$$

pinches the center-line $t = 1/2$ of $A^0_{p,0}$ to $p = (0, 0)$, and when restricted to the half-open annuli $A^0_{p,\delta} - \{t = 1/2\}$, it becames n_j-fold and n_i-fold cyclic coverings over the punctured banks $D_2 - \{p\}$ and $D_1 - \{p\}$, respectively.

Identifying D_2 and D_1 with the unit disk $\{z \mid |z| \leq 1\}$ through the parametrizations $z \mapsto (0, \varepsilon z)$ and $z \mapsto (\varepsilon z, 0)$, define $T_2 : D_2 \to D_2$ and $T_1 : D_1 \to D_1$ by setting

$$T_2(z) = ze\left(\frac{\pi}{m_i}(1 - |z|)\right)$$

and

$$T_1(z) = ze\left(\frac{\pi}{m_j}(1 - |z|)\right).$$

Recall that the monodromy homeomorphism $f = h_{2\pi} : F_\delta \to F_\delta$ was defined, within Δ_p, as follows:

$$f(z_1, z_2) = \left(e\left(\frac{2\pi(1-t)}{m_j}\right)z_1, e\left(\frac{2\pi t}{m_i}\right)z_2\right),$$

where $t = t(|z_1|, |z_2|)$.

Then it is easy to verify

$$T_2\pi = \pi f \quad \text{on} \quad A_{p,\delta} \cap \left\{t \leq \frac{1}{2}\right\}$$

and

$$T_1\pi = \pi f \quad \text{on} \quad A_{p,\delta} \cap \left\{t \geq \frac{1}{2}\right\}.$$

Thus if we define the pinched covering $\pi : F_\delta \to F_0$ to be the union $\bigcup_p (\pi|_{A_{p,\delta}}) \cup \bigcup_i (\pi|_{B_{i,\delta}})$, π satisfies condition (v) in the definition of a generalized quotient given in Chap. 3. (Note that π is well-defined because $\pi|_{A_{p,\delta}}$ and $\pi|_{B_{i,\delta}}$ coincide on the intersection $A_{p,\delta} \cap B_{i\delta}$ whenever it is non-empty.) The other conditions (i)–(iv) in the same definition are readily checked on $\pi : F_\delta \to F_0$ and $f : F_\delta \to F_\delta$, using the results of this Chap. 7.

This completes the proof of Theorem 7.4. □

Chapter 8
Blowing Down Is a Topological Operation

The notation will be the same as in the previous chapter. The following fact is well-known. Suppose F_0 has only normal crossings.

Proposition 8.1. *Let Θ_0 be an irreducible component of F_0 with multiplicity m_0. Let $\{p_1, p_2, \ldots, p_k\}$ be the set of intersection points between Θ_0 and the other irreducible components of F_0. Let m_i be the multiplicity of the irreducible component which intersects Θ_0 at p_i $(i = 1, 2, \ldots, k)$. Then $m_1 + m_2 + \cdots + m_k$ is divisible by m_0, and the quotient $(m_1 + m_2 + \cdots + m_k)/m_0$ is equal to $-\Theta_0 \cdot \Theta_0$, where $\Theta_0 \cdot \Theta_0$ denotes the self-intersection number of Θ_0 as a 2-cycle in M.*

Proof. By the definition of the multiplicities, it is clear that a general fiber is homologous in M to the divisor-expression of F_0, $m_0\Theta_0 + \sum_{j\neq 0} m_j\Theta_j$. Since Θ_0 does not intersect a general fiber, we have

$$\begin{aligned} 0 &= \Theta_0 \cdot \left(m_0\Theta_0 + \sum_{j\neq 0} m_j\Theta_j \right) \\ &= m_0\Theta_0 \cdot \Theta_0 + (m_1 + m_2 + \cdots + m_k). \end{aligned}$$

The proposition follows now easily from this. □

We will call an irreducible component Θ a (-1)-*curve* if it is a smoothly embedded 2-sphere with $\Theta \cdot \Theta = -1$.

Definition 8.1. F_0 is said to be *normally minimal* if F_0 has only normal crossings and every irreducible component Θ which is a (-1)-curve intersects the other irreducible components in more than two points. (If F_0 is normally minimal, we cannot blow down any irreducible component without producing a singular point which is not a normal crossing).

Corollary 8.1. *Suppose F_0 has only normal crossings. F_0 is normally minimal if and only if the following conditions are satisfied*

Y. Matsumoto and J.M. Montesinos-Amilibia, *Pseudo-periodic Maps and Degeneration of Riemann Surfaces*, Lecture Notes in Mathematics 2030,
DOI 10.1007/978-3-642-22534-5_8,

(i) *If an irreducible component* Θ_0 *is an embedded sphere and intersects only one other component* Θ_i *in one point exactly, then the multiplicity* m_0 *of* Θ_0 *divides the multiplicity* m_i *of* Θ_i *and* $m_i/m_0 \geq 2$.

(ii) *If an irreducible component* Θ_0 *is an embedded sphere and intersects the other components only in two points* $\{p_1, p_2\}$*, then the multiplicity* m_0 *of* Θ_0 *divides* m_1+m_2 *and* $(m_1+m_2)/m_0 \geq 2$*, where* m_i *is the multiplicity of the irreducible component* Θ_i *intersecting* Θ_0 *at* p_i, $i = 1, 2$.

Remark 8.1. Corollary 8.1 motivated the definition of a minimal quotient given in Chap. 3.

Essentially the same argument as in the proof of Proposition 8.1 can be done if F_0 is not assumed to have only normal crossings. In this general case, irreducible components and multiplicities are defined as follows (Cf. [32]). A point p of F_0 is called a *generic point of multiplicy* m if there exist coordinates (z_1, z_2) around p such that $z_1(p) = z_2(p) = 0$ and $\psi(z_1, z_2) = (z_2)^m$. It is easy to see that the multiplicity is locally constant, thus constant on a connected component of the set of generic points of F_0. The closure of such a connected component is called an *irreducible component* of F_0. The meaning of the *multiplicity* of an irreducible component would be evident.

Let $F_0 = m_1\Theta_1 + m_2\Theta_2 + \cdots + m_s\Theta_s$ be the expression of F_0 as a divisor (in the general case). Then by the same reason as in the case of normal crossings, we have the following equalities

$$\Theta_i \cdot (m_1\Theta_1 + m_2\Theta_2 + \cdots + m_s\Theta_s) = 0, \quad i = 1, 2, \ldots, s. \qquad (*)$$

Using $(*)$, we can prove

Proposition 8.2. *Assume the general fiber* F_ξ *has genus* ≥ 1. *Let* Θ_i, $i = 1, 2$, *be the irreducible components of* F_0 *which are* (-1)*-curves. Then* Θ_1 *and* Θ_2 *do not intersect*: $\Theta_1 \cap \Theta_2 = \emptyset$.

Proof. Note that if $\Theta_i \cap \Theta_j \neq \emptyset$ and $i \neq j$, then $\Theta_i \cdot \Theta_j > 0$, where Θ_i and Θ_j are any two irreducible components of F_0 (see [65, Sect. 20], [45]). We will show that the assumption $\Theta_1 \cap \Theta_2 \neq \emptyset$ leads to contradiction.

Case 1. $\Theta_1 \cdot \Theta_2 > 1$.
Putting $i = 1$ in $(*)$, we have

$$0 = -m_1 + m_2\Theta_1 \cdot \Theta_2 + m_3\Theta_1 \cdot \Theta_3 + \cdots > -m_1 + m_2. \qquad (A)$$

Putting $i = 2$, we have

$$0 = m_1\Theta_2 \cdot \Theta_1 - m_2 + m_3\Theta_2 \cdot \Theta_3 + \cdots > m_1 - m_2. \qquad (B)$$

Obviously (A) contradicts (B).

Case 2. $\Theta_1 \cdot \Theta_2 = 1$.
Putting $i = 1$ in $(*)$, we have

$$0 = -m_1 + m_2 + m_3\Theta_1 \cdot \Theta_3 + \cdots \geq -m_1 + m_2. \qquad (A')$$

in which the equality holds iff Θ_1 does not intersect Θ_i, $i \geq 3$.
Putting $i = 2$, we have

$$0 = m_1 - m_2 + m_3\Theta_2 \cdot \Theta_3 + \cdots \geq m_1 - m_2. \qquad (B')$$

in which the equality holds iff Θ_2 does not intersect Θ_i, $i \geq 3$.

Clearly (A') contradicts (B') unless the equalities hold in (A') and (B'). Therefore $m_1 = m_2$ and, since F_0 is connected, F_0 consists of Θ_1 and Θ_2 only. i.e. the shape of F_0 is very restricted:

$$F_0 = m\Theta_1 + m\Theta_2, \quad \Theta_1 \cdot \Theta_2 = 1.$$

Blow down Θ_2, then $F_0 = m\Theta_1'$, where Θ_1' is a smoothly embedded 2-sphere with $\Theta_1' \cdot \Theta_1' = 0$. This implies that the general fiber F_ξ is a disjoint union of m 2-spheres, which contradicts the assumption of Proposition 8.2.

This completes the proof of Proposition 8.2. □

Let $\hat{\mathscr{S}}_g$ denote the set of all topological types of degenerating families (M, D, ψ) of genus g in which the central fibers F_0 are *normally minimal*. Let

$$\hat{\rho} : \hat{\mathscr{S}}_g \to \mathscr{P}_g^-$$

denote the map sending (M, D, ψ) to the topological monodromy around $F_0 = \psi^{-1}(0)$.

Recall that we defined in Chap. 7 a similar set $\mathscr{S}_g$ consisting of all topological types of *minimal* degenerating families of genus g and a similar map $\rho : \mathscr{S}_g \to \mathscr{P}_g^-$.

The purpose of the present chapter is to prove

Theorem 8.1. *There exists an onto map $\beta : \hat{\mathscr{S}}_g \to \mathscr{S}_g$ such that $\rho \circ \beta = \widehat{\rho}$.*

Remark 8.2. $\beta : \hat{\mathscr{S}}_g \to \mathscr{S}_g$ is given by blowing down (-1)-curves.

Proof of 8.1. Let (M, D, ψ) be a degenerating family of genus g. If there are (-1)-curves in (M, D, ψ), we blow down one of them. Generally speaking, there still remain (-1)-curves, and/or some irreducible components of $\psi^{-1}(0)$ may presently change into (-1)-curves after blowing down. Then we blow down one of these (-1)-curves. Repeating this process finitely many times, we get a minimal degenerating family (M_*, D, ψ_*) of genus g.

Lemma 8.1. *(M_*, D, ψ_*) is independent of the ordered sequence of the (-1)-curves according to which they are blown down.*

Proof. We call the (-1)-curves in (M, D, ψ) the (-1)-curves of *the* 1-*st generation*. They are disjoint by Proposition 8.2. If we blow down them simultaneously, some irreducible components may change into (-1)-curves. We call them the (-1)-curves *of the* 2-*nd generation*. They are disjoint. If we blow down them simultaneously, some irreducible components may again change into (-1)-curves, which we call the (-1)-curves *of the 3-rd generation*, and so on. This process terminates after finitely many steps, and we get a minimal degenerating family (M_{st}, D, ψ_{st}). We call it the *standard minimal degenerating family* obtained from (M, D, ψ). It will be shown that (M_*, D, ψ_*) is nothing but (M_{st}, D, ψ_{st}).

Let

$$(\Theta_1,\ \Theta_2,\ \ldots,\ \Theta_k)$$

be the ordered sequence of (appropriately re-indexed) (-1)-curves such that (M_*, D, ψ_*) is obtained from (M, D, ψ) by blowing down Θ_i after Θ_{i-1} for $i = 2, \ldots, k$. The first Θ_1 must belong to the 1-st generation. Suppose Θ_j be the (-1)-curve in the sequence which belongs to the 1-st generation and is the nearest to Θ_1 in the sense of the sequence. Then by Proposition 8.2, any (-1)-curve in the sequence between Θ_1 and Θ_j is disjoint from Θ_j. Thus we can change the position of Θ_j to the one next to Θ_1 without affecting the final result, so that the sequence

$$(\Theta_1,\ \Theta_j,\ \Theta_2,\ \ldots,\ \Theta_{j-1},\ \Theta_{j+1},\ \ldots,\ \Theta_k)$$

gives the same (M_*, D, ψ_*).

Repeating this argument and re-indexing the irreducible components if necessary, we get a new sequence giving (M_*, D, ψ_*),

$$(\Theta_1,\ \Theta_2,\ \ldots,\ \Theta_k),$$

in which there exists an index $i(1)$ such that

$$\Theta_1,\ \Theta_2,\ \ldots,\ \Theta_{i(1)}$$

are (-1)-curves of the 1-st generation, while

$$\Theta_{i(1)+1},\ \Theta_{i(1)+2},\ \ldots,\ \Theta_k$$

are not. Note that there are left in M no (-1)-curves of the 1-st generation other than

$$\Theta_1,\ \Theta_2,\ \ldots,\ \Theta_{i(1)}$$

because if there were any, (M_*, D, ψ_*) would contain the (-1)-curves of the 1-st generation. This contradicts the minimality of (M_*, D, ψ_*). Thus

$$\Theta_1,\ \Theta_2,\ \ldots,\ \Theta_{i(1)}$$

all are (-1)-curves of the 1-st generation.

After blowing down $\Theta_1, \ldots, \Theta_{i(1)}$, we get a new degenerating family $(M_{(1)}, D, \psi_{(1)})$ in which (-1)-curves of the 2-nd generation may appear in general. In that case, the $(i(1)+1)$-th curve $\Theta_{i(1)+1}$ must be of the second generation, and we can repeat the same argument in the manifold $M_{(1)}$ as in M.

Proceeding in this way, we get a sequence giving (M_*, D, ψ_*),

$$(\Theta_1, \Theta_2, \ldots, \Theta_{i(1)}, \Theta_{i(1)+1}, \ldots, \Theta_{i(2)}, \ldots, \Theta_{i(n)})$$

in which

$$\Theta_1, \Theta_2, \ldots, \Theta_{i(1)}$$

are of the first generation, and

$$\Theta_{i(l-1)+1}, \Theta_{i(l-1)+2}, \ldots, \Theta_{i(l)}$$

all are (-1)-curves of the l-th generation, for $l = 2, \ldots, n$. This implies that (M_*, D, ψ_*) is nothing but $(M_{\mathrm{st}}, D, \psi_{\mathrm{st}})$, as asserted. □

By Lemma 8.1 we may call (M_*, D, ψ_*) *the minimal degenerating family* obtained from (M, D, ψ).

Now let (M, D, ψ) and (M', D', ψ') be two degenerating families of genus g which are topologically equivalent. The point in proving Theorem 8.1 is to show that the minimal degenerating families (M_*, D, ψ_*) and (M'_*, D', ψ'_*) obtained from (M, D, ψ) and (M', D', ψ'), respectively, are also topologically equivalent.

To prove this we need two more lemmas and some corollaries.

To state Lemma 8.2, let us recall the Milnor fibering from [45]. Let p be a point of $F_0 = \psi^{-1}(0)$. We consider ψ as a holomorphic function defined around p, Let S_ε be a 3-sphere of radius $\varepsilon > 0$ centered at p, where ε is sufficiently small. The *Milnor fibering* of ψ at p is a fibering over S^1 defined by

$$\frac{\psi}{|\psi|} : S_\varepsilon - K_\varepsilon \to S^1,$$

where $K_\varepsilon = S_\varepsilon \cap \psi^{-1}(0)$. Its typical fiber is called the *Milnor fiber*.

Let $\Lambda : M \to M'$ and $\lambda : D \to D'$ be orientation-preserving homeomorphisms such that $\lambda(0) = 0$ and $\psi'\Lambda = \lambda\psi$.

Lemma 8.2. *The Milnor fiber of ψ at p is homeomorphic to the Milnor fiber of ψ' at $p' = \Lambda(p)$.*

Proof. Let D_ε be the 4-ball centered at p of radius ε. If the complex number c is sufficiently close to 0, the interior of $F_{\varepsilon,c} = D_\varepsilon \cap \psi^{-1}(c)$ is homeomorphic to the Milnor fiber of ψ at p (see [45]). Suppose ε and ε' are two small numbers with $0 < \varepsilon' < \varepsilon$. If $|c|$ is small enough, $(D_\varepsilon - IntD_{\varepsilon'}) \cap \psi^{-1}(c)$ is homeomorphic to $(\delta F_{\varepsilon,c}) \times [\varepsilon', \varepsilon]$, which is a disjoint union of annuli. Cf. [37].

We take likewise a small 4-ball Δ_δ centered at p' of radius $\delta > 0$. Suppose δ, ε and ε' are chosen so that

$$D_\varepsilon \Subset \Lambda^{-1}(\Delta_\delta) \Subset D_{\varepsilon'}$$

where $A \Subset B$ means $A \subset Int(B)$. See Fig. 8.1.

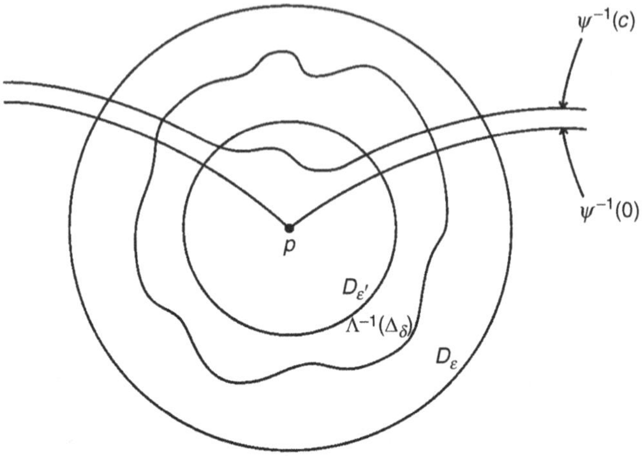

Fig. 8.1 Topological invariance of Milnor fibers

Setting $F'_{\delta,\lambda(c)} = \Delta_\delta \cap (\psi')^{-1}(\lambda(c))$, we see that each connected component (= an annulus) of $(D_\varepsilon - Int(D_{\varepsilon'})) \cap \psi^{-1}(c)$ contains exactly one component (= a circle) of $\Lambda^{-1}(\partial F'_{\delta,\lambda(c)})$ and that the two boundary components of the annulus are separated by the component of $\Lambda^{-1}(\partial F'_{\delta,\lambda(c)})$. Then it follows, by the annulus theorem, that each component of $F_{\varepsilon,c} - \Lambda^{-1}(Int F'_{\delta,\lambda(c)})$ is homeomorphic to an annulus. Thus $F_{\varepsilon,c}$ is homeomorphic to $F'_{\delta,\lambda(c)}$, whose interior is homeomorphic to the Milnor fiber of ψ' at p' if δ is sufficiently small. This completes the proof of Lemma 8.2 □

Corollary 8.2. *If p ($\in F_0$) is a generic point of multiplicity m, then $p' = \Lambda(p)$ ($\in F'_0$) is a generic point of the same multiplicity.*

Proof. By the definition of a generic point (given after Corollary 8.1), there exist local coordinates (z_1, z_2) centered at p such that $\psi(z_1, z_2) = (z_2)^m$ around p.

Now decompose ψ' (which is considered as a holomorphic function around $p' = \Lambda(p)$) into irreducible factors:

$$\psi' = f_1^{m_1} f_2^{m_2} \cdots f_r^{m_r}, \quad \text{with} \quad f_i(p') = 0, \quad i = 1, \ldots, r,$$

where r is equal to the number of the branches of $F'_0 = (\psi')^{-1}(0)$ passing through p'. This number r is a topological invariant. (This follows from the fact that a 3-sphere of a small radius centered at p' intersects F'_0 transversely in a link of r components. See [45].) Note that only one branch of $F_0 = \psi^{-1}(0)$, namely $z_2 = 0$, passes through p and that Λ maps (F_0, p) homeomorphically onto (F'_0, p'). Thus $r = 1$ and we may assume

$$\psi' = (f_1)^{m_1},$$

where f_1 is irreducible with $f_1(p') = 0$.

The Milnor fiber of ψ' at p' is a disjoint union of m_1 copies of the (connected) Milnor fiber of f' at p', while the Milnor fiber of ψ at p is clearly a disjoint union of m open 2-disks. Thus by Lemma 8.2, $m_1 = m$ and the Milnor fiber of f_1 at p' is an open 2-disk. It follows that f_1 is non-singular at p' (see [45]), which proves that p' is a generic point of multiplicity m. □

Recall that an irreducible component of $F_0 = \psi^{-1}(0)$ is the closure of a connected component of the set of generic points of F_0. The following corollary follows from Corollary 8.2.

Corollary 8.3. *If Θ is an irreducible component of F_0, then $\Lambda(\Theta)$ is an irreducible component of F_0'. The multiplicitics of Θ and $\Lambda(\Theta)$ are equal.*

F_0 is non-singular if and only if every point of F_0 is a generic point of multiplicity 1. Thus we have

Corollary 8.4. *If $F_0 = \psi^{-1}(0)$ is non-singular and if (M, D, ψ) is topologically equivalent to (M', D', ψ'), then $F_0' = (\psi')^{-1}(0)$ is non-singular.*

The next lemma will give a topological characterization of a (-1)-curve. Working for a while in a more general setting, let Θ be a compact complex irreducible curve (namely, a closed Riemann surface which may be singular) in a complex manifold M of complex dimension 2. Let $p_1, p_2, \dots, p_k$ be the singular points of Θ, and let D_i be a small 4-ball centered at p_i. Let T be a tubular neighborhood of $\Theta - \bigcup_{i=1}^k (\Theta \cap IntD_i)$ in $M - \bigcup_{i=1}^k IntD_i$. We will call $N = T \cup \left(\bigcup_{i=1}^k D_i\right)$ an *admissible neighborhood* of Θ. (See Fig. 8.2.)

It can be proved that $N - \Theta$ is homeomorphic to $\partial N \times [0, 1)$. See [45].

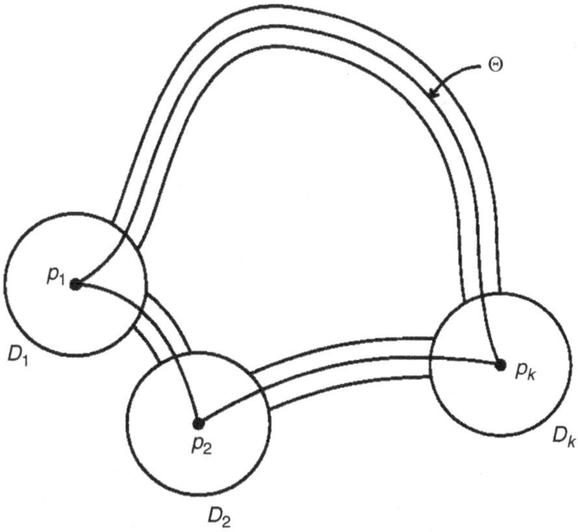

Fig. 8.2 Admissible neighborhood of Θ

Lemma 8.3. *An irreducible curve Θ is a smoothly embedded 2 -sphere with $\Theta \cdot \Theta = \pm 1$ if and only if $N - \Theta$ is simple connected.*

Proof. The "only if" part is trivial. Conversely we will prove that if $N - \Theta$ is simply connected, then Θ is a smoothly embedded 2-sphere with $\Theta \cdot \Theta = \pm 1$. Since $N - \Theta \cong \partial N \times [0, 1)$, the assumption is equivalent to the simple connectivity of ∂N.

If Θ has self-intersection points or has positive genus, neither Θ nor ∂N is simply connected. Thus if ∂N is simply connected, Θ is a topologically embedded 2-sphere. Let $p_1, p_2, \ldots, p_k$ be the singular points of Θ, and let D_i, $i = 1, \ldots, k$, be small 4-balls, each centered at p_i. Let T be a tubular neighborhood of $\check{\Theta}$ $\left(= \Theta - \bigcup_{i=1}^{k} \Theta \cap \mathit{IntD}_i\right)$ in $M - \bigcup_{i=1}^{k} \mathit{IntD}_i$, as before. Let E_i denote the complement $\partial D_i - \mathit{Int}(\partial D_i \cap T)$, for $i = 1, \ldots, k$. Then we have $\partial N = \partial T \cup \bigcup_{i=1}^{k} E_i$.

Case 1. $k \geq 2$. ∂T is an orientable S^1-bundle over $\check{\Theta}$ (= a sphere with k holes $\simeq$ a bouquet of $k - 1$ S^1 s). Thus $\partial T \cong \check{\Theta} \times S^1$, whose identification will be fixed throughout. Firstly, we will see that the inclusion

$$\partial E_1 = \partial T \cap E_1 \to \partial T$$

induces an injective homomorphism $\pi_1(\partial E_1) \to \pi_1(\partial T)$ between fundamental groups. In fact, ∂E_1 is identified with (a boundary component of $\check{\Theta}$) $\times S^1$, and since $k \geq 2$, any non-zero multiple of any boundary component of $\check{\Theta}$ is not null-homotopic in $\check{\Theta}$. Secondly we note that the knot $\partial D_1 \cap \Theta$ is non-trivial in ∂D_1 because p_1 is a singular point [45]. Thus the inclusion $\partial E_1 \to E_1$ also induces an injective homomorphisms $\pi_1(\partial E_1) \to \pi_1(E_1)$ [54]. By Van Kampen's theorem and [36, Theorem 4.3], the homomorphisms

$$\pi_i(\partial T) \to \pi_1(\partial T \cup E_1)$$

and

$$\pi_i(E_1) \to \pi_1(\partial T \cup E_1)$$

are injective. We can repeat the above argument inductively to show that

$$\pi_1(\partial T \cup E_1 \cup \cdots \cup E_{i-1}) \to \pi_1(\partial T \cup E_1 \cup \cdots \cup E_i)$$

and

$$\pi_i(E_i) \to \pi_1(\partial T \cup E_1 \cup \cdots \cup E_i)$$

are injective, for $i = 1, 2, \ldots, k$. In particular, $\pi_1(\partial N)$ is non-trivial. This contradicts the assumption $\pi_1(\partial N) \cong \{1\}$.

Case 2. $k = 1$. The manifold $\partial N = \partial T \cup E_1$ is obtained from the 3-sphere ∂D_1 by Dehn-surgery along the knot $K = \partial D_1 \cap \Theta$. By the Burau and Kähler theorem, K is a (non-trivial) iterated torus knot [45], which has property P, [15]. Thus $\pi_1(\partial N) \ncong \{1\}$, contradicting the assumption.

So far we have proved that if ∂N is simply connected, then $k = 0$, i.e. Θ has no singular points and is a smoothly embedded 2-sphere. In this case, it holds that $\pi_1(\partial N) \cong \mathbf{Z}/l$, where $l = |\Theta \cdot \Theta|$. Thus if ∂N is simply connected, we have $\Theta \cdot \Theta = \pm 1$. This completes the proof of Lemma 8.3. □

Corollary 8.5. *If Θ is a (-1)-curve, then so is $\Lambda(\Theta)$.*

The proof is left to the reader.

Remark 8.3. If Θ is a (-1)-curve, the boundary ∂N of the admissible neighborhood is diffeomorphic to 3-sphere.

Completion of the proof of Theorem 8.1

Let (M, D, ψ) and (M', D', ψ') be degenerating families of genus $g \geq 1$. Assuming that they are topologically equivalent under orientation-preserving homeomorphisms

$$(\Lambda, \lambda) : (M, D) \to (M', D')$$

such that $\lambda(0) = 0$ and $\psi' \Lambda = \lambda \psi$, we will prove that the minimal degenerating families (M_*, D, ψ_*) and (M'_*, D', ψ'_*) obtained from (M, D, ψ) and (M', D', ψ'), respectively, are topologically equivalent.

Let

$$(\Theta_1, \Theta_2, \ldots, \Theta_k)$$

be an ordered sequence of (-1)-curves giving (M_*, D, ψ_*). The operation of blowing down Θ_1 is equivalent to the following one: delete Θ_1 from M and compactify the created end by adding one point. By Corollary 8.5, $\Lambda(\Theta_1)$ is a (-1)-curve in M'. Thus, by this interpretation, blowing down Θ_1 and $\Lambda(\Theta_1)$ in M and M', respectively, gives topologically equivalent degenerating families $(M_{(1)}, D, \psi_{(1)})$ and $(M'_{(1)}, D', \psi'_{(1)})$. Repeating this argument for Θ_i, $i = 1, 2, \ldots, k$, we can conclude that (M_*, D, ψ_*) is topologically equivalent to $(M'_{(k)}, D', \psi'_{(k)})$ which is obtained from (M', D', ψ') by blowing down $\Lambda(\Theta_i)$, for $i = 1, 2, \ldots, k$ in this order. By Corollary 8.5, $(M'_{(k)}, D', \psi'_{(k)})$ is minimal, and by Lemma 8.1, $(M'_{(k)}, D', \psi'_{(k)})$ is nothing but (M'_*, D', ψ'_*). Thus (M_*, D, ψ_*) is topologically equivalent to (M'_*, D', ψ'_*) as asserted.

We have completed the proof that for $g \geq 1$ the map $\beta : \hat{\mathscr{S}}_g \to \mathscr{S}_g$ which sends $[M, D, \psi]$ to $[M_*, D, \psi_*]$ is well-defined. Moreover it is *onto* because any minimal degenerating family (M_*, D, ψ_*) can be blown up to an (M, D, ψ) whose central fiber has only normal crossings, [33].

Finally, for $g \geq 2$, the diagram

$$\begin{array}{ccc}
\hat{\mathscr{S}}_g & \xrightarrow{\hat{\rho}} & \mathscr{P}_g^- \\
{\scriptstyle \beta}\downarrow & & \downarrow{\scriptstyle =} \\
\mathscr{S}_g & \xrightarrow[\rho]{} & \mathscr{P}_g^- .
\end{array}$$

commutes because blowing down does not affect the topological monodromy around the central fiber. This completes the proof of Theorem 8.1. □

Remark 8.4. For $g = 1$, we have the commutative diagram:

$$\begin{array}{ccc} \hat{\mathscr{S}}_1 & \xrightarrow{\hat{\rho}} & \{\text{conjugacy classes in } SL(2,\mathbf{Z})\} \\ \beta\downarrow & & \downarrow = \\ \mathscr{S}_1 & \xrightarrow[\rho]{} & \{\text{conjugacy classes in } SL(2,\mathbf{Z})\}. \end{array}$$

Chapter 9
Singular Open-Book

In this chapter, we will complete the proof of Theorem 7.2. We wish to show that the monodromy correspondence

$$\rho : \mathscr{S}_g \to \mathscr{P}_g^-$$

is bijective for $g \geq 2$. In Chap. 8, we proved the existence of an *onto* map $\beta : \hat{\mathscr{S}}_g \to \mathscr{S}_g$ such that the diagram commutes (Theorem 8.1):

$$\begin{array}{ccc} \hat{\mathscr{S}}_g & \xrightarrow{\hat{\rho}} & \mathscr{P}_g^- \\ {\scriptstyle\beta}\downarrow & & \downarrow{\scriptstyle=} \\ \mathscr{S}_g & \xrightarrow[\rho]{} & \mathscr{P}_g^-. \end{array}$$

Clearly, the desired bijectivity of ρ will follow from that of $\hat{\rho}$.

To prove that $\hat{\rho}$ is bijective, we take a pseudo-periodic map of negative twist,

$$f : \Sigma_g \to \Sigma_g,$$

and will construct a normally minimal degenerating family of genus g,

$$(M,\ D,\ \psi)$$

whose topological monodromy coincides with f up to isotopy and conjugation. This will give a well-defined map

$$\hat{\sigma} : \mathscr{P}_g^- \to \hat{\mathscr{S}}_g,$$

which will be the inverse of $\hat{\rho}$.

Y. Matsumoto and J.M. Montesinos-Amilibia, *Pseudo-periodic Maps and Degeneration of Riemann Surfaces*, Lecture Notes in Mathematics 2030,
DOI 10.1007/978-3-642-22534-5_9,

The construction proceeds in two steps:

The 1-st step (mapping cylinder).

We may assume that the pseudo-periodic map $f : \Sigma_g \to \Sigma_g$ is in superstandard form and has the minimal quotient $\pi : \Sigma_g \to S$, (Theorem 3.1). Take the disjoint union $\Sigma_g \times [0, 1] \cup S$ and identify each $(y, 0) \in \Sigma_g \times [0, 1]$ with $\pi(y) \in S$. The resulting quotient space C_π is the *mapping cylinder* of π. C_π is not in general a manifold.

The 2-nd step (Singular open-book).

The notion of open-book decomposition of a manifold was introduced by Winkelnkemper [69], and independently by Tamura [61] under the terminology of "spinnable structure". Our construction here is very similar to theirs, except that the "binding" is a chorizo space instead of a manifold. The "page" is the mapping cylinder C_π, and the characteristic map $\tilde{f} : C_\pi \to C_\pi$ is the trace of a certain deformation of $f : \Sigma_g \to \Sigma_g$, which we will now describe.

Recall that for each node p of S, we fixed a closed nodal neighborhood $\overline{N_p}$ (Chap. 3), and that both banks of $\overline{N_p}$ were parametrized so that they were identified with $\{z \mid |z| \le 1\}$, (see the definition of generalized quotient in Chap. 3). Now we take an isotopy

$$I_r : S \to S$$

parametrized by r $(0 < r \le 1)$ such that (i) $I_1 = id_S$, and (ii) for each closed nodal neighborhood $\overline{N_p}$ and for each bank $\overline{D_p}$ of $\overline{N_p}$, parametrized as $\{z \mid |z| \le 1\}$, we have $(I_r \mid \overline{D_p})(z) = rz$.

Using $\{I_r\}_{0<r\le 1}$, define an isotopy

$$\tilde{I}_r : \Sigma_g \to \Sigma_g$$

likewise parametrized by r $(0 < r \le 1)$ such that

(i) $\tilde{I}_1 = id_{\Sigma_g}$, and

(ii) $\pi\, \tilde{I}_r = I_r\, \pi$,

where $\pi : \Sigma_g \to S$ is the pinched covering.

The mentioned deformation of f is defined to be

$$f_r := \tilde{I}_r\, f\, (\tilde{I}_r)^{-1}, \quad 0 < r \le 1.$$

Note that $f_1 = f$ and that f_r commutes with $\pi : \Sigma_g \to S$ over $S_r := S - \bigcup_{p=\text{node}} I_r(N_p)$, i.e. induces the identity of S_r. (N.B. By the definition of a generalized quotient given in Chap. 3, f is a collection of covering translations over $S - \bigcup_{p=\text{node}} N_p$.) Thus as $r \to 0$, this induced map converges to the identity of S. (But the "limit" of f_r, as $r \to 0$, fails to be continuous along $\bigcup_{p=\text{node}} \pi^{-1}(p)$.)

The definition of the characteristic map $\tilde{f} : C_\pi \to C_\pi$ is the following:

$$\tilde{f}(c) = \begin{cases} (f_r(y), r) & \text{if } \ c = (y, r) \in \Sigma_g \times [0, 1], \ r > 0 \\ c & \text{if } \ c \in S. \end{cases}$$

As we remarked, the limit of f_r as $r \to 0$ is not continuous along $\bigcup_{p=\text{node}} \pi^{-1}(p)$, but $\pi^{-1}(p)$ collapses to a point in S. Thus this discontinuity causes no problem and $\tilde{f} : C_\pi \to C_\pi$ is a homeomorphism satisfying $\tilde{f}|_S = id_S$.

Our singular open-book $\overline{M}$ is constructed first by taking the mapping torus of $\tilde{f} : C_\pi \to C_\pi$ and then collapsing the "sub-torus" corresponding to $id_S = \tilde{f}|_S : S \to S$ onto S.

More precisely, we set

$$\overline{M} = [0,\ 2\pi] \times C_\pi / \sim,$$

where the equivalence relation $\sim$ is generated by

$$(2\pi,\ c) \sim (0,\ \tilde{f}(c)) \quad \text{for} \quad \forall c \in C_\pi,$$

and

$$(\theta,\ c) \sim (0,\ c) \quad \text{for} \quad \forall c \in S,\ \forall \theta \in [0, 2\pi].$$

This completes the 2-nd step of our construction.

It will be convenient to represent a point of $\overline{M}$ by $[\theta,\ y,\ r]$ ($\theta \in [0, 2\pi]$, $y \in \Sigma_g$, $r \in [0, 1]$) under the equivalence relation generated by

(i) $[\theta,\ y,\ 0] = [\theta,\ y',\ 0]$ iff $\pi(y) = \pi(y')$,
(ii) $[2\pi,\ y,\ r] = [0,\ f_r(y),\ r]$ for $r > 0$,
(iii) $[\theta,\ y,\ 0] = [0,\ y,\ 0]$.

S can be identified with a subspace of $\overline{M}$ by the correspondence:

$$S \ni \pi(y) \longleftrightarrow [0,\ y,\ 0] \in \overline{M}.$$

Define $\partial\overline{M}$ to be $\{[\theta,\ y,\ r] \in \overline{M} \mid r = 1\}$, and M to be $\overline{M} - \partial\overline{M}$. Also define a map $\overline{\psi} : \overline{M} \to \mathbf{C}$ by

$$\overline{\psi}([\theta,\ y,\ r]) = r\ \exp(\sqrt{-1}\theta).$$

We set $\psi = \overline{\psi}|_M$.

Theorem 9.1. *M admits a structure of complex manifold of complex dimension two such that $\psi : M \to D$ is a surjective proper holomorphic map satisfying*

(i) $\psi^{-1}(0) = S$,
(ii) $\psi|_{M^*} : M^* \to D^*$ *is a smooth fiber bundle with fiber* Σ_g, *where* $M^* = M - \psi^{-1}(0)$ *and* $D^* = D - \{0\}$, *and*
(iii) the topological monodromy coincides with f *up to isotopy and conjugation.*

Proof. We shall construct for each point $p \in S$ a small complex coordinate neighborhood W_p ($\subset M$) containing p. There are two cases to be considered:

Case A. p is a node.

Case B. p is a generic point.

First we consider Case A.

Suppose p is a node, and is an intersection point of two irreducible components, say Θ_1 and Θ_2, of S. Let m_1 and m_2 be the multiplicities of Θ_1 and Θ_2, respectively. Set $m = \gcd(m_1, m_2)$ and $n_1 = m_1/m, n_2 = m_2/m$.

Remember that a closed nodal neighborhood $\overline{N_p}$ of p has been fixed (Chap. 3).

We define a subspace $\overline{W_p}$ of $\overline{M}$ by

$$\overline{W_p} = \{[\theta,\ y,\ r] \mid \pi(y) \in N_p\},$$

where N_p is the (open) nodal neighborhood corresponding to $\overline{N_p}$ and $\pi : \Sigma_g \to S$ is the pinched covering.

Set

$$\overline{W_p}(\theta_0) = \{[\theta,\ y,\ r] \in \overline{W_p} \mid \theta = \theta_0 \text{ (const.)}\},$$

then $\overline{W_p}(0)$ is identified with $C_\pi \cap \overline{W_p}$.

By the definition of generalized quotient (Chap. 3), the outer-most face of $\overline{W_p}(0)$, i.e.

$$\partial\overline{W_p}(0) := \{[0,\ y,\ r] \in \overline{W_p} \mid r = 1\},$$

consists of m annuli,

$$A_1,\ A_2,\ \ldots,\ A_m,$$

which are permuted by f cyclically:

$$f(A_\alpha) = A_{\alpha+1}, \quad \alpha = 1,\ 2,\ \ldots,\ m-1,$$

and $f(A_m) = A_1$. The m-th power

$$f^m : A_\alpha \to A_\alpha$$

is a linear twist with the screw number $-1/n_1 n_2$. Also by the definition of generalized quotient (Chap. 3), there exist parametrizations of the banks of N_p,

$$\{z \mid |z| < 1\} \to D_1,$$

$$\{z \mid |z| < 1\} \to D_2,$$

such that if we identify D_i with $\{z \mid |z| < 1\}$ through these parametrizations ($i = 1, 2$) and define

$$T_i : D_i \to D_i$$

($i = 1, 2$) by

$$T_1(z) = z \exp\left(\frac{\sqrt{-1}\pi}{m_2}(1 - |z|)\right),$$

$$T_2(z) = z \exp\left(\frac{\sqrt{-1}\pi}{m_1}(1 - |z|)\right),$$

then the following identities hold:

$$T_i \pi = \pi f \quad \text{on} \quad \pi^{-1}(D_i) \cap \partial\overline{W_p}(0), \ \ i = 1, 2.$$

Consider the unit (open) polydisk $Int\Delta$ in $\mathbf{C}^2$:

$$Int\Delta = \{(z_1, z_2) \mid |z_1| < 1, \ |z_2| < 1\}.$$

Let ξ be a complex number (with $0 \leq |\xi| < 1$) and let A_ξ denote the fiber of

$$(z_1)^{m_2}(z_2)^{m_1} : Int\Delta \to \mathbf{C}$$

over ξ:

$$A_\xi = \{(z_1, z_2) \in Int\Delta \mid (z_1)^{m_2}(z_2)^{m_1} = \xi\}.$$

Set $e(\theta) = \exp(\sqrt{-1}\theta)$ and take a small positive real δ. As we did in Chap. 7, we define a continuous family of homeomorphisms

$$h_\theta : A_\delta \to A_{e(\theta)\delta}, \quad 0 \leq \theta \leq 2\pi$$

as follows:

$$h_\theta(z_1, z_2) = \left(e\left(\frac{(1-t)\theta}{m_2}\right) z_1, \ e\left(\frac{t\theta}{m_1}\right) z_2\right),$$

where $t = t(|z_1|, |z_2|)$ is the real-valued function $t : Int\Delta \to [0, 1]$ introduced in Chap. 7 (See Fig. 7.3):

$$t = \begin{cases} \dfrac{1 - |z_2|}{2(1 - |z_1|)} & \text{if} \quad |z_1| \leq |z_2| < 1, \\[2ex] 1 - \dfrac{1 - |z_1|}{2(1 - |z_2|)} & \text{if} \quad 1 > |z_1| \geq |z_2|. \end{cases}$$

(Note that ε in the definition of t in Chap. 7 is here replaced by 1). The final stage of h_θ is the monodromy

$$f' = h_{2\pi} : A_\delta \to A_\delta$$

of

$$(z_1)^{m_2}(z_2)^{m_1} : Int\Delta \to \mathbf{C}.$$

As we proved in Chap. 7, A_δ consists of m annuli, permuted cyclically by $f' : A_\delta \to A_\delta$, and on each annulus, the m-th power of f' is a linear twist with the screw number $-1/n_1 n_2$. (Theorem 7.3). Moreover, let

$$\pi'|_{A_\delta} : A_\delta \to A_0$$

denote the restriction, to A_δ, of the map

$$\pi' : Int\Delta \to A_0$$

defined in the proof of Theorem 7.4:

$$\pi'(z_1, z_2) = \begin{cases} \left(0, \frac{|z_2|-|z_1|}{1-|z_1|} \cdot \frac{z_2}{|z_2|}\right) \in D'_2 & \text{if } |z_1| \le |z_2| < 1, \\ \left(\frac{|z_1|-|z_2|}{1-|z_2|} \cdot \frac{z_1}{|z_1|}, 0\right) \in D'_1 & \text{if } 1 > |z_1| \ge |z_2|. \end{cases}$$

Here the ε in Chap. 7 is replaced again by 1, and the two banks of A_0 are denoted by D'_1 and D'_2 ($\pi'|_{A_\delta} : A_\delta \to A_0$ is a pinched covering). If we identify these banks with $\{z \mid |z| < 1\}$ through the parametrizations $z \mapsto (z, 0)$ and $z \mapsto (0, z)$, then we have

$$(\pi'|_{A_\delta})\, f' = T_1\, (\pi'|_{A_\delta}) \quad \text{on} \quad (\pi')^{-1}(D'_1) \cap A_\delta$$

and

$$(\pi'|_{A_\delta})\, f' = T_2\, (\pi'|_{A_\delta}) \quad \text{on} \quad (\pi')^{-1}(D'_2) \cap A_\delta.$$

(See the proof of Theorem 7.4).

Therefore, there exist homeomorphisms

$$H\,|_{\partial\overline{W_p}(0)} : \partial\overline{W_p}(0) \to A_\delta$$

and

$$H|_{N_p} : N_p \to A_0$$

satisfying

1. $\pi'\,(H|_{\partial\overline{W_p}(0)}) = (H|_{N_p})\,\pi$,
2. $f'\,(H|_{\partial\overline{W_p}(0)}) = (H|_{\partial\overline{W_p}(0)})\,f$, and
3. $T_i\,(H|_{N_p}) = (H|_{N_p})\,T_i$, $i = 1, 2$.

Next, for each r with $0 \le r \le 1$, let

$$\pi'_r : A_\delta \to A_{r\delta}$$

be the projection which satisfies the following conditions:

If $(z_1, z_2) \in A_\delta$ and $\pi'_r(z_1, z_2) = (z_1^{(r)}, z_2^{(r)})$, then

(a) $t(|z_1|, |z_2|) = t(|z_1^{(r)}|, |z_2^{(r)}|)$,

(b) $|(z_1^{(r)})^{m_2}(z_2^{(r)})^{m_1}| = r\delta$, and

(c) $\arg(z_1) = \arg(z_1^{(r)})$, $\arg(z_2) = \arg(z_2^{(r)})$.

Note that these conditions uniquely determine the projection

$$\pi'_r : A_\delta \to A_{r\delta}.$$

If $0 < r \leq 1$,

$$\pi'_r : A_\delta \to A_{r\delta}$$

is a homeomorphism (in particular, π'_1 = identity of A_δ), and if $r = 0$,

$$\pi'_0 : A_\delta \to A_0$$

coincides with

$$\pi'|_{A_\delta} : A_\delta \to A_0.$$

It can be easily checked that

$$(\pi'|_{A_{r\delta}})\, \pi'_r = (\pi'|_{A_\delta}).$$

We set

$$\overline{V} = \{(z_1, z_2) \in Int\Delta \mid |(z_1)^{m_2}(z_2)^{m_1}| \leq \delta\}$$

and define $\overline{V}(\theta_0)$ by

$$\overline{V}(\theta_0) = \{(z_1, z_2) \in \overline{V} \mid \arg(z_1)^{m_2}(z_2)^{m_1} = \theta_0\} \cup A_0.$$

Let us extend

$$(H|_{\partial\overline{W_p}(0)}) \cup (H|_{N_p}) : \partial\overline{W_p}(0) \cup N_p \to A_\delta \cup A_0$$

to

$$H|_{\overline{W_p}(0)} : \overline{W_p}(0) \to \overline{V}(0)$$

by

$$H([0,\ y,\ r]) := \pi'_r\,(H([0,\ y,\ 1])),$$

where $H([0,\ y,\ 1]) = (H|_{\partial\overline{W_p}(0)})(y)$. This $H|_{\overline{W_p}(0)}$ is a homeomorphism:

$$\overline{W_p}(0) \to \overline{V}(0).$$

Let

$$\tau(t, r) : [0, 1] \times (0, 1] \to [0, 1]$$

be a real-valued function defined by

$$\tau(t, r) = \begin{cases} 0 & 0 \leq t \leq \frac{1-r}{2}, \\ \frac{1}{r}\left(t - \frac{1-r}{2}\right) & \frac{1-r}{2} \leq t \leq \frac{1+r}{2}, \\ 1 & \frac{1+r}{2} \leq t \leq 1. \end{cases}$$

Using this function, we define a homeomorphism

$$\tilde{h}_\theta : \overline{V}(0) \to \overline{V}(\theta)$$

as follows:

If $(z_1, z_2) \notin A_0$, set

$$\tilde{h}_\theta(z_1, z_2) = \left(e\left(\frac{\theta}{m_2}(1 - \tau(t, r))\right) z_1,\ e\left(\frac{\theta}{m_1}\tau(t, r)\right) z_2 \right),$$

where $t = t(|z_1|, |z_2|)$ and $r = |(z_1)^{m_2}(z_2)^{m_1}|/\delta > 0$, and if $(z_1, z_2) \in A_0$, set

$$\tilde{h}_\theta(z_1, z_2) = (z_1, z_2).$$

It is not difficult to see that $\tilde{h}_\theta$ is in fact a homeomorphism.

We are now in a position to define the desired homeomorphism

$$H : \overline{W_p} \to \overline{V}$$

which extends

$$H|_{\overline{W_p}(0)} : \overline{W}_p(0) \to \overline{V}(0)$$

and gives complex coordinates to $W_p (= \overline{W_p} - \partial \overline{W_p})$. The definition is as follows:

$$H([\theta,\ y,\ r]) := \tilde{h}_\theta\, H([0,\ y,\ r]).$$

To check the well-definedness of H, we will use a lemma, in which we identify each bank D_i' ($i = 1$ or 2) of A_0 with $D = \{z \mid |z| < 1\}$, and let

$$I_r : D \to D$$

denote the map $I_r(z) = rz$.

Lemma 9.1. *Let (z_1, z_2) be a point of $A_{r\delta}$, and put $z = \pi'(z_1, z_2) \in A_0 = D_1' \cup D_2'$. If $z \in D_i'$ ($i = 1$ or 2), then*

$$\pi' \tilde{h}_{2\pi}(z_1, z_2) = \begin{cases} z & \text{if } \ r \leq |z| \leq 1, \\ I_r\, T_i\, I_r^{-1}(z) & \text{if } \ 0 \leq |z| \leq r. \end{cases}$$

Proof. Suppose $z \in D_1'$. Then $t = t(|z_1|, |z_2|) \geq 1/2$, and by the definition of $\pi'(z_1, z_2)$ we can check $|z| = 2t - 1$. We have

$$\pi' \tilde{h}_{2\pi}(z_1, z_2) = z\, e\left(\frac{2\pi}{m_2}(1 - \tau(t,\ r))\right)$$

$$= \begin{cases} z & \text{if} \quad \dfrac{1+r}{2} \le t \le 1, \\ z\, e\left(\dfrac{2\pi}{m_2}\left(\dfrac{1}{2} - \dfrac{1}{r}\left(t - \dfrac{1}{2}\right)\right)\right) & \text{if} \quad \dfrac{1}{2} \le t \le \dfrac{1+r}{2}, \end{cases}$$

$$= \begin{cases} z & \text{if} \quad r \le |z| \le 1, \\ z\, e\left(\dfrac{\pi}{m_2}\left(1 - \dfrac{|z|}{r}\right)\right) & \text{if} \quad 0 \le |z| \le r, \end{cases}$$

$$= \begin{cases} z & \text{if} \quad r \le |z| \le 1, \\ I_r\, T_1\, I_r^{-1}(z) & \text{if} \quad 0 \le |z| \le r. \end{cases}$$

When $z \in D_2'$, the proof is similar. □

Now let us prove the well-definedness of H. It will be sufficient to check it on each of the three generating relations for $[\theta,\ y,\ r]$:

(i) $H([\theta,\ y,\ 0]) = H([\theta,\ y',\ 0])$ *if* $\pi(y) = \pi(y')$.
In fact,

$$H([\theta,\ y,\ 0]) = \tilde{h}_\theta\, H([0,\ y,\ 0]) = \tilde{h}_\theta\, \pi_0'\, (H|_{\partial \overline{W_p(0)}})(y)$$
$$= \tilde{h}_\theta\, (H|_{N_p})\pi(y) = (H|_{N_p})\, \pi(y),$$

because $(H|_{N_p})\, \pi(y) \in A_0$. Thus (i) follows.

(ii) $H([2\pi,\ y,\ r]) = H([0,\ f_r(y),\ r])$ *for* $r > 0$.
First we will prove

$$\pi'\, H([2\pi,\ y,\ r]) = \pi'\, H([0,\ f_r(y),\ r]) \quad (\in A_0). \tag{$*$}$$

In fact, noting that

$$\pi'\, \pi_r'\, H([0,\ y,\ 1]) = \pi'(H|_{\partial \overline{W_p(0)}}) = (H|_{N_p})\, \pi(y),$$

and using Lemma 9.1, we have

$$\pi'\, H([2\pi,\ y,\ r]) = \pi'\, \tilde{h}_{2\pi}\, H([0, y, r])$$
$$= \pi'\, \tilde{h}_{2\pi}\, \pi_r'(H[0, y, 1])$$
$$= \begin{cases} (H|_{N_p})\, \pi(y) & \text{if } r \le |(H|_{N_p})\, \pi(y)| \le 1 \\ I_r\, T_i\, I_r^{-1}\, (H|_{N_p})\, \pi(y),\ (i{=}1 \text{ or } 2) & \text{if } 0 \le |(H|_{N_p})\, \pi(y)| \le r \end{cases}$$
$$= \begin{cases} (H|_{N_p})\, \pi(y) \\ (H|_{N_p})\, I_r\, T_i\, I_r^{-1}\, \pi(y) \end{cases}$$

$$= (H|_{N_p})\, \pi\, f_r(y) = \pi'\, (H|_{\partial \overline{W_p}(0)})\, f_r(y)$$
$$= \pi'\, \pi_r'\, (H|_{\partial \overline{W_p}(0)})\, f_r(y) = \pi'\, H([0,\ f_r(y), r]).$$

Thus $(*)$ is verified.

Note that $(\pi'|A_{r\delta})\ :\ A_{r\delta}\ \to\ A_0$ is a covering map except over one point $(0, 0)$, and that the correspondences $y \mapsto H([2\pi,\ y,\ r])$ and $y \mapsto H([0,\ f_r(y),\ r])$ give two homeomorphisms $\partial \overline{W_p}(0) \to A_{r\delta}$ which coincide when $r = 1$;

$$H([2\pi,\ y,\ 1]) = \widetilde{h}_{2\pi}\, H[0,\ y,\ 1] = f'\,(H|_{\partial \overline{W_p}(0)})(y)$$
$$= (H|_{\partial \overline{W_p}(0)})\, f(y) = H([0,\ f(y),\ 1]).$$

Moreover, these homeomorphisms continuously change, depending on $r > 0$. Therefore, the identity $(*)$ implies in fact that

$$H([2\pi,\ y,\ r]) = H([0,\ f_r(y),\ r])$$

for $r\ (0 < r \leq 1)$. This proves (ii).

(iii) $H([\theta,\ y,\ 0]) = H([0,\ y,\ 0])$.

This is trivial because both sides are equal to $(H|_{N_p})\, \pi(y)$. (See the proof of (i)).

This completes the proof of well-definedness of $H : \overline{W_p} \to \overline{V}$.

The proof that H is a homeomorphism is left to the reader.

By the definition of $\psi : W_p \to \mathbf{C}$,

$$\psi([\theta,\ y,\ r]) = r\, e(\theta).$$

On the other hand, if

$$H([\theta,\ y,\ r]) = (z_1,\ z_2),$$

then

$$|(z_1)^{m_2}(z_2)^{m_1}| = r\delta$$

because $H([\theta,\ y,\ r]) \in A_{r\delta}$, and also we have

$$\arg(z_1)^{m_2}(z_2)^{m_1} = \theta$$

because $H([\theta,\ y,\ r]) = \tilde{h}_\theta\, H([0,\ y,\ r]) \in \overline{V}(\theta)$.

Thus if we define $z_1' = \delta' z_1$ and $z_2' = \delta' z_2$, where $\delta' = (\delta)^{-1/(m_1+m_2)}$, then with these coordinates $(z_1',\ z_2')$ on Δ, we have the commutative diagram:

$$\begin{array}{ccc} W_p & \xrightarrow{H} & Int\Delta \\ \psi\downarrow & & \downarrow (z_1')^{m_2}(z_2')^{m_1} \\ \mathbf{C} & \xrightarrow[=]{} & \mathbf{C}. \end{array}$$

The complex coordinates (z_1', z_2') pulled back to W_p by H are the desired ones.

This completes the (somewhat lengthy) consideration of Case A in the proof of Theorem 9.1.

Next, we consider Case B where $p \in S$ is a generic point of S. Suppose p is on an irreducible component Θ_1 whose multiplicity is $m_1 \geq 1$. Let U be a sufficiently small open disk-neighborhood of p in Θ_1. We may assume that U does not contain any node of S and intersects at most one closed nodal neighborhood $\overline{N_{p'}}$ in S.

For a small positive real $r_0 > 0$, define $\overline{W_p}^{(r_0)} (\subset \overline{M})$ by

$$\overline{W_p}^{(r_0)} = \{[\theta, y, r] \mid \pi(y) \in U, \ r \leq r_0\}.$$

The boundary $\partial\overline{W_p}^{(r_0)} (= \{[\theta, y, r] \in \overline{W_p}^{(r_0)} \mid r = r_0\})$ consists of m_1 copies of U, denoted by $\tilde{U}_1, \tilde{U}_2, \ldots, \tilde{U}_m$, and $f : \Sigma_g \to \Sigma_g$ permutes them cyclically. If r_0 is sufficiently small, $\overline{W_p}^{(r_0)}$ is disjoint from the "twisting region" of $\overline{W_{p'}}$ and $f^{m_1}|_{\tilde{U}_\alpha} : \tilde{U}_\alpha \to \tilde{U}_\alpha$ is the identity ($\alpha = 1, 2, \ldots, m$). See Fig. 9.1. From this, it follows that $W_p^{(r_0)} (= \overline{W_p}^{(r_0)} - \partial\overline{W_p}^{(r_0)})$ is homeomorphic to an open 4-disk if r_0 is small enough.

In Case A we constructed complex coordinates (z_1', z_2') in $W_{p'}$ such that $\psi|_{W_{p'}} = (z_1')^{m_2}(z_2')^{m_1}$. (We assume that the bank of $N_{p'}$ on the irreducible component Θ_1 is given by $z_2' = 0$.) Since Θ_1 is a closed oriented surface, we can put a complex structure on Θ_1, and may assume z_1' gives the complex coordinate on $W_{p'} \cap \Theta_1$. Let z_1'' be a complex coordinate on U that is compatible with the complex structure of Θ_1. The coordinate change

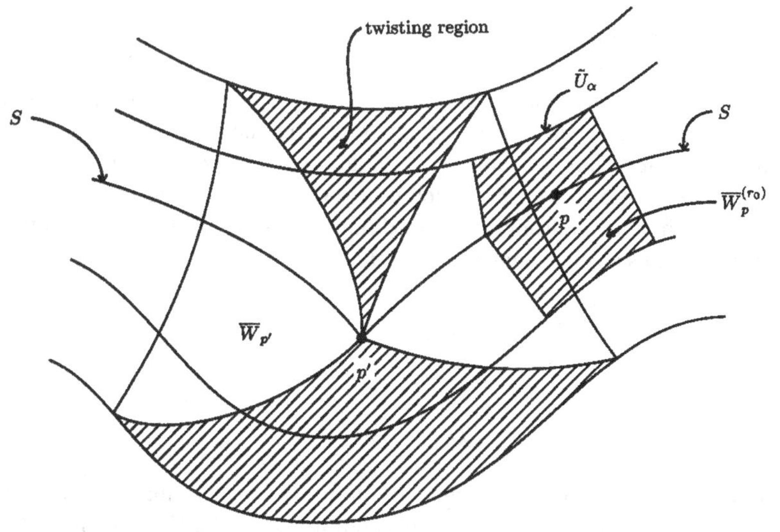

Fig. 9.1 Construction of complex coordinates

$$z_1'' = z_1''(z_1') \quad \text{on} \quad U \cap W_{p'}$$

is bi-holomorphic.

Let us define complex coordinates (z_1'', z_2'') on $W_p^{(r_0)}$ as follows:

If $q \in (W_p^{(r_0)} - W_{p'})$, then put

$$(z_1''(q), z_2''(q)) = (z_1''(\pi(q)), \psi(q)^{1/m_1}),$$

and if $q \in W_{p'} \cap W_p^{(r_0)}$, then put

$$(z_1''(q), z_2''(q)) = (z_1''(z_1', z_2'), \psi(q)^{1/m_1}),$$

where $z_1''(z_1', z_2')$ stands for the composition

$$q = (z_1', z_2') \mapsto (z_1', 0) \mapsto z_1' \mapsto z_1''(z_1') \qquad (*)$$

where the first arrow stands for "projection" and the third arrow stands for "coordinate change".

Since the composition $(*)$ converges to the projection $\pi : q \mapsto \pi(q)$, as $\pi(q)$ $(\in N_{p'} \cap \Theta_1)$ approaches the boundary of the bank $N_{p'} \cap \Theta_1$, the above two cases continuously match up to give complex coordinates (z_1'', z_2'') on $W_p^{(r_0)}$. The coordinate change between (z_1', z_2') (on $W_{p'}$) and (z_1'', z_2'') (on $W_p^{(r_0)}$) is easily seen to be biholomorphic.

The function $\psi|_{W_p}$ is given by

$$\psi = (z_2'')^{m_1}$$

in terms of the coordinates (z_1'', z_2'').

Case B is completed.

In this way we have constructed complex coordinates on a neighborhood of each point of S. The coordinate changes among them are easily checked to be biholomorphic. Thus we have constructed a complex structure on an open neighborhood W of S in M with which $\psi|_W : W \to \mathbf{C}$ is holomorphic.

Evidently, there is a (shrinking) embedding $h : M \to W$ into W such that the diagram commutes (with a small real $\eta > 0$):

$$\begin{array}{ccc} M & \xrightarrow{h:\hookrightarrow} & W \\ \psi \big\downarrow & & \big\downarrow (\psi \mid W) \\ D & \xrightarrow[\times \eta]{} & D. \end{array}$$

Pulling back the complex structure of W to M, we can make M a complex manifold of complex dimension two. The conditions (i), (ii) and (iii) in the statement of Theorem 9.1 are readily verified. This completes the proof of Theorem 9.1. □

9.1 Completion of the Proof of Theorem 7.2

We will define a map

$$\hat{\sigma} : \mathscr{P}_g^- \to \hat{\mathscr{S}}_g,$$

which is to be the inverse to the monodromy correspondence

$$\hat{\rho} : \hat{\mathscr{S}}_g \to \mathscr{P}_g^-.$$

Let $\langle f \rangle$ be any element of $\mathscr{P}_g^-$. By Theorem 3.1,

$$f : \Sigma_g \to \Sigma_g$$

is isotopic to an

$$f' : \Sigma_g \to \Sigma_g$$

in superstandard form. Let

$$\pi : \Sigma_g \to S$$

be its minimal quotient. By the singular open-book construction starting from f' and π, we obtain a degenerating family (M_f, D_f, ψ_f) whose monodromy coincides with f up to isotopy and conjugation (Theorem 9.1). This (M_f, D_f, ψ_f) is normally minimal because

$$\pi : \Sigma_g \to S$$

is the minimal quotient (Corollary 8.1).

Suppose $\langle f_1 \rangle = \langle f_2 \rangle$, i.e. f_1 coincides with f_2 up to isotopy and conjugation. Then by Theorem 4.2, there exist a homeomorphism

$$h : \Sigma_g \to \Sigma_g$$

and a numerical homeomorphism

$$H : S[f_1] \to S[f_2]$$

such that the diagram commutes:

$$\begin{array}{ccc} \Sigma_g & \xrightarrow{h} & \Sigma_g \\ {\scriptstyle \pi_1}\downarrow & & \downarrow{\scriptstyle \pi_2} \\ S[f_1] & \xrightarrow[H]{} & S[f_2] \end{array}$$

In other words, their minimal quotients

$$\pi_1 : \Sigma_g \to S[f_1]$$

and

$$\pi_2 : \Sigma_g \to S[f_2]$$

are topologically equivalent. If f_1' and f_2' are the corresponding superstandard forms, then by Theorem 4.2 (iii),

$$f_1' = h^{-1} f_2' h.$$

Since the construction of a singular open-book made in this chapter is purely topological and has no topological ambiguity once the initial data $f' : \Sigma_g \to \Sigma_g$ (in superstandard form) and $\pi : \Sigma_g \to S$ are given, we conclude that the degenerating families (M_1, D_1, ψ_1) and (M_2, D_2, ψ_2) resulting from f_1' and f_2' respectively are topologically equivalent. Thus sending $\langle f \rangle \in \mathscr{P}_g^-$ to the topological equivalence class $[M_f, D_f, \psi_f]$ of (M_f, D_f, ψ_f), we obtain a well-defined map

$$\hat{\sigma} : \mathscr{P}_g^- \to \hat{\mathscr{S}}_g.$$

Lemma 9.2. $\hat{\rho}\,\hat{\sigma} = id.$

This is clear because the topological monodromy of (M_f, D_f, ψ_f) is equal to $\langle f \rangle$ (Theorem 9.1).

To prove the bijectivity of $\hat{\rho}$, it remains to show

Lemma 9.3. $\hat{\sigma}\,\hat{\rho} = id.$

Proof. Let $[M, D, \psi]$ be any element of $\hat{\mathscr{S}}_g$. By Theorem 7.4, the monodromy homeomorphism

$$f : F_\delta \to F_\delta,$$

constructed in Chap. 7, is a pseudo-periodic map in superstandard form and there exist a pinched covering

$$\pi : F_\delta \to F_0$$

which is a generalized quotient of f. This

$$\pi : F_\delta \to F_0$$

is actually the minimal quotient because F_0 is normally minimal (cf. Corollary 8.1.). By essentially the same argument as was done in the proof of Theorem 9.1 (especially in Case A), M is shown to have the structure of a singular open-book (minus the boundary) which is constructed starting from the data

$$f : F_\delta \to F_\delta$$

and

$$\pi : F_\delta \to F_0.$$

Therefore, $[M, D, \psi]$ coincides with

$$[M_f, D_f, \psi_f] = \hat{\sigma}(\langle f \rangle) = \hat{\sigma}\,\hat{\rho}([M, D, \psi]).$$

This proves $\hat{\sigma}\,\hat{\rho} = id$. □

As we remarked at the beginning of Chap. 9, the bijectivity of $\hat{\rho} : \hat{\mathscr{S}}_g \to \mathscr{P}_g^-$ (just proved) implies the bijectivity of $\rho : \mathscr{S}_g \to \mathscr{P}_g^-$. Thus Theorem 7.2 is completely proved. □

Recall the diagram

$$\begin{array}{ccc} \hat{\mathscr{S}}_g & \xrightarrow{\hat{\rho}} & \mathscr{P}_g^- \\ {\scriptstyle \beta}\downarrow & & \downarrow{\scriptstyle =} \\ \mathscr{S}_g & \xrightarrow[\rho]{} & \mathscr{P}_g^- \end{array}$$

from Chap. 8. Now we have proved that $\hat{\rho}$ and ρ are bijective maps. This readily implies

Proposition 9.1. *The map $\beta : \hat{\mathscr{S}}_g \to \mathscr{S}_g$ is bijective.*

9.2 Characterization of the Triples (S, Y, c) That Come from Pseudo-periodic Maps

Let S be a connected numerical chorizo space which satisfies the minimality condition (Chap. 4); Y the decomposition diagram of S (Chap. 6), namely, a weighted graph whose vertices (resp. edges) are in one-to-one correspondence to the bodies (resp. arches) of S. Let $c \in H^1_W(Y)$ be a class in the weighted cohomology group of Y (Chap. 6). In this paragraph, we will give a necessary and sufficient condition under which the triple (S, Y, c) comes from a pseudo-periodic map $f : \Sigma_g \to \Sigma_g$ of negative twist, of a surface of a given genus $g \geq 2$.

There are several necessary conditions for this that we have already proved:

1. Let Θ_0 be any irreducible component of S, $\{p_1, p_2, \ldots, p_k\}$ the set of the intersection points among Θ_0 and the other irreducible components. Let m_i be the multiplicity of the irreducible component which intersects Θ_0 at p_i $(i = 1, 2, \ldots, k)$, and let m_0 be the multiplicity of Θ_0. Then m_0 divides $m_1 + m_2 + \cdots + m_k$ (see Proposition 3.1).
2. Let e be an edge of Y. Then the weight of e is equal to the gcd of the successive multiplicities on the arch corresponding to e (see Corollary 6.2. (i)).
3. Let v be a vertex of Y. Suppose the core part P_0 of the body corresponding to v is contained in an irreducible component Θ_0. Let m_0 be the multiplicity of Θ_0. Let $m_1, m_2, \ldots, m_k$ be the multiplicities that have the same meaning as in (1).

   If P_0 has genus 0, then the weight of v is equal to gcd$(m_0, m_1, \ldots, m_k)$. If P_0 has genus ≥ 1, then the weight of v divides gcd$(m_0, m_1, \ldots, m_k)$ (see Corollary 6.2 (ii) and (iii).)

4. Let W_0 be the gcd of the weights of all vertices of Y. Then the homomorphism

$$c_* : H_1(Y : \mathbf{Z}) \to \mathbf{Z}/W_0$$

determined by $c \in H^1_W(Y)$ is onto. (See Lemma 6.6.)

5. Let $S = m_1\Theta_1 + m_2\Theta_2 + \cdots + m_s\Theta_s$ be the expression of S as a (formal) divisor. Let P_i be a part of Θ_i (cf. Chap. 3). Then

$$2 - 2g = \sum_{i=1}^{s} m_i \chi(P_i) \quad (< 0)$$

where $\chi(P_i)$ is the Euler characteristic of P_i.

Theorem 9.2. *Given a triple (S, Y, c) satisfying the above conditions (1) $\sim$ (5), there exists a pseudo-periodic map $f : \Sigma_g \to \Sigma_g$ of negative twist whose minimal quotient $S[f]$ and its decomposition diagram are topologically equivalent to S and Y in such a way that the cohomology classes $c[f]$ and c naturally correspond. More precisely, let $\eta_f : S[f] \to Y[f]$ and $\eta : S \to Y$ denote the collapsing maps. Then there exist a numerical homeomorphism $H : S[f] \to S$ and a weighted isomorphism $\Psi : Y[f] \to Y$ such that the diagram*

$$\begin{array}{ccc} S[f] & \xrightarrow{\eta_f} & Y[f] \\ {\scriptstyle H}\downarrow & & \downarrow{\scriptstyle \Psi} \\ S & \xrightarrow[\eta]{} & Y \end{array}$$

commutes and $\Psi^*(c) = c[f]$.

The proof is indirect in the sense that it applies our results of Part II on degenerating families of Riemann surfaces of genus g. We will start with Winters' existence theorem restated in our context.

Theorem 9.3 (Winters [70, Corollary 4.3]). *Let*

$$S = m_1\Theta_1 + m_2\Theta_2 + \cdots + m_s\Theta_s$$

be a numerical chorizo space satisfying the above condition (1). Then there exist a complex surface M and a proper holomorphic map

$$\psi : M \to D$$

onto an open unit disk D such that $\psi^{-1}(0)$ has normal crossings and such that

$$m_1\Theta_1 + m_2\Theta_2 + \cdots + m_s\Theta_s$$

is the divisor expression of $\psi^{-1}(0)$.

Outline of the proof. From topological viewpoint, Winters' argument may be interpreted as follows:

First one embeds Θ_i (considered as a Riemann surface) in a certain complex surface N_i. In case Θ_i has no self-intersection points, the description of N_i is easy; N_i is a holomorphic bundle over Θ_i whose fiber is biholomorphically homeomorphic to $\mathbf{C}$ and which admits a cross section (this is the embedded image of Θ_i). The self-intersection number of Θ_i in N_i is

$$-\frac{1}{m_i}\left(\sum_{j\neq i} m_j\Theta_j \cdot \Theta_i\right),$$

where $\Theta_j \cdot \Theta_i$ denotes the number of the nodes of S which are contained in $\Theta_i \cap \Theta_j$. The above number is an integer because of Condition (1). Also this condition assures the existence of such an N_i that admits a holomorphic map $\psi_i : N_i \to \mathbf{C}$ whose zero-set $\psi_i^{-1}(0)$ (as a divisor) is expressed as

$$m_i\Theta_i + \sum_{k=1}^{r(i)} m_k^{(i)} c_k^{(i)},$$

where $c_1^{(i)}, c_2^{(i)}, \ldots, c_{r(i)}^{(i)}$ are the fibers of the bundle $N_i \to \Theta_i$ over the intersection points, $p_1^{(i)}, p_2^{(i)}, \ldots, p_{r(i)}^{(i)}$, of Θ_i with the other irreducible components, $m_k^{(i)}$ being the multiplicity of the irreducible component which meets Θ_i at $p_k^{(i)}$ ($k = 1, 2, \ldots, r(i)$). See Fig. 9.2.

In case Θ_i has self-intersection points, N_i is an immersed image of a holomorphic $\mathbf{C}$-bundle $\tilde{N}_i$ over the normalized $\tilde{\Theta}_i$ ($= \Theta_i$ with the self-intersections separated). $\tilde{N}_i$ admits a cross section whose self-intersection number is equal to

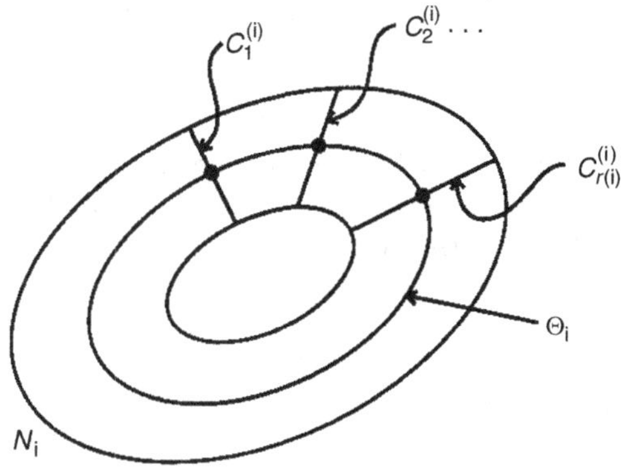

Fig. 9.2 Other components intersect Θ_i transversely

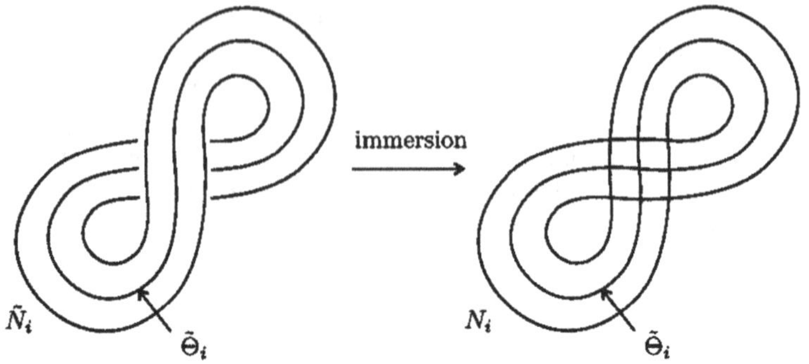

Fig. 9.3 Θ_i has self-intersections

$$-\frac{1}{m_i}\left(\sum_{j\neq i} m_j\Theta_j\cdot\Theta_i\right)-2K_i,$$

K_i denoting the number of the self-intersection points of Θ_i. The reason for the appearance of $-2K_i$ is that each (transverse) self-intersection point of Θ_i contributes 2 to the self-intersection number $\Theta_i\cdot\Theta_i$ in N_i. (See Fig. 9.3)

Similarly to the case of a non-singular Θ_i, one can adjust the construction of the N_i so that it has a holomorphic map $\psi_i : N_i \to \mathbf{C}$ whose divisor $\psi_i^{-1}(0)$ consists of $m_i\Theta_i$ and $m_1^{(i)}c_1^{(i)}+\cdots+m_{r(i)}^{(i)}c_{r(i)}^{(i)}$, where $c_1^{(i)},\ldots,c_{r(i)}^{(i)}$ have the same meaning as before. At a self-intersection point of Θ_i, ψ_i is locally expressed as

$$\psi_i=(z_1z_2)^{m_i}$$

with appropiate coordinates (z_1,z_2). Now we have obtained the parts (one for each Θ_i) to construct $\psi : M\to D$.

To get M, one has only to do "plumbing" with these parts

$$N_1,\ N_2,\ \ldots,\ N_s,$$

i.e. glue N_i and N_j at each intersection point of Θ_i and Θ_j (in S) so that the fiber and the base coordinates are interchanged at the time of gluing.

At an intersection point of Θ_i and Θ_j,

$$\psi_i : N_i\to\mathbf{C}$$

is locally expressed as

$$\psi_i=(z_1)^{m_j}(z_2)^{m_i}$$

with appropiate coordinates (z_1,z_2). Thus one can make the plumbing so that ψ_i's are patched together to give a holomorphic map $\psi : M\to\mathbf{C}$. We can easily modify

M and ψ to obtain a surjective, proper and holomorphic map $\psi : M \to D$ whose fiber over 0, $\psi^{-1}(0)$, coincides with the numerical chorizo space S (as a divisor).

This completes the outline of the proof of Theorem 9.3. □

Addendum 1 *In the above construction, we may assume that the number of the connected components of the general fiber of* $\psi_i : N_i \to C$ *is equal to*

$$\gcd(m_i,\ m_1^{(i)},\ m_2^{(i)},\ \ldots,\ m_{r(i)}^{(i)}),$$

for each $i = 1, 2, \ldots, s$.

N.B. The general fiber of $\psi_i : N_i \to \mathbf{C}$ is non-compact if $r(i) \geq 1$.

Proof. Let Θ_0 be any irreducible component of S; m_0 its multiplicity. Suppose Θ_0 intersects the other irreducible components in r points

$$p_1,\ p_2,\ \ldots,\ p_r.$$

Let

$$m_1,\ m_2,\ \ldots,\ m_r$$

be the multiplicities of the irreducible components which intersect Θ_0 in

$$p_1,\ p_2,\ \ldots,\ p_r,$$

respectively.

First suppose Θ_0 has no self-intersection points, in which case, N_0 can be constructed as follows.

Take an open disk D_0 on Θ_0 having coordinate z_1 and containing

$$p_1,\ p_2,\ \ldots,\ p_r.$$

The values of the z_1-coordinate for

$$p_1,\ p_2,\ \ldots,\ p_r$$

in D_0 are supposed to be

$$a_1,\ a_2,\ \ldots,\ a_r.$$

Let D_1 be another open disk on Θ_0 such that $D_1 \supset \overline{D_0}$.

Suppose $\mathbf{C}$ has coordinate z_2. The bundle N_0 is constructed by holomorphically pasting $D_1 \times \mathbf{C}$ and $(\Theta_0 - \overline{D_0}) \times \mathbf{C}$ along $(D_1 - \overline{D_0}) \times \mathbf{C}$ so that the holomorphic maps

$$\psi' = (z_1 - a_1)^{m_1}(z_1 - a_2)^{m_2} \cdots (z_1 - a_r)^{m_r}(z_2)^{m_0} : D_1 \times \mathbf{C} \to \mathbf{C}$$

and

$$\psi'' = (z_2)^{m_0} : (\Theta_0 - \overline{D_0}) \times \mathbf{C} \to \mathbf{C}$$

coincide on $(D_1 - \overline{D_0}) \times \mathbf{C}$. (The condition that m_0 divides

$$m_1 + m_2 + \cdots + m_r$$

assures that the fibers of

$$\psi' : D_1 \times \mathbf{C} \to \mathbf{C}$$

make a trivial foliation in $(D_1 - \overline{D_0}) \times \mathbf{C}$, each leaf covering $D_1 - \overline{D_0}$ just once.)

If N_0 is so constructed, the general fiber of the resulting holomorphic map

$$\psi_0 \, (= \psi' \cup \psi'') : N_0 \to \mathbf{C}$$

has

$$\gcd(m_0, m_1, \ldots, m_r)$$

number of components. (cf. Proposition 4.2.)

This completes the proof when Θ_0 has no self-intersection points. If it has self-intersection points, the proof is essentially the same. □

9.3 Completion of the Proof of Theorem 9.2

Note that the general fiber of $\psi : M \to D$ constructed by Theorem 9.3 is not necessarily connected. But just as in the proof of Theorem 7.4, we can construct a pinched covering

$$\pi : F_\delta \to F_0 = S,$$

from the general fiber $F_\delta = \psi^{-1}(\delta)$ to the central fiber $F_0 = \psi^{-1}(0)$ (identified with S), where δ ($\in D$) is a small positive real.

Let $ARCH_0$ be an arch of S. Then by the argument before Lemma 4.2, $\pi^{-1}(ARCH_0)$ consists of m annuli, m being the gcd of the multiplicities of the successive irreducible components on $ARCH_0$. By Condition (2), this number m is equal to the weight of the edge $\eta(ARCH_0)$ of Y.

Let BDY_0 be a body of S. Let P_0 be the core part of BDY_0, Θ_0 the irreducible component containing P_0. Let

$$m_0, m_1, \ldots, m_k$$

have the same meaning as in Condition (3). By Addendum 1, we may assume that the number of the connected components of $\pi^{-1}(BDY_0)$ is equal to

$$\gcd(m_0, m_1, \ldots, m_k).$$

If P_0 has genus 0, this number coincides with the weight of the vertex $\eta(BDY_0)$ of Y. (Condition (3).) If P_0 has genus ≥ 1, the weight of the vertex $\eta(BDY_0)$ divides

$$\gcd(m_0, m_1, \ldots, m_k).$$

We will change M and ψ so that the number of the connected components of $\pi^{-1}(BDY_0)$ is equal to the weight, say w, of $\eta(BDY_0)$:

We can take a pair of simple closed curves C_1 and C_2 on P_0 cutting each other transversely once, because P_0 has genus ≥ 1. Cut open $N_0(\subset M)$ along $C_1 \times \mathbf{C}$ (which is the bundle over C_1 induced from the $\mathbf{C}$-bundle $N_0 \to \Theta_0$) then reglue the two copies $(C_1 \times \mathbf{C})_0$ and $(C_1 \times \mathbf{C})_1$ by the following rotation of the fibers:

$$(z_1, z_2) \mapsto (z_1, \exp(2\pi w\sqrt{-1}/m_0)z_2), \quad (z_1, z_2) \in C_1 \times \mathbf{C}.$$

Since w divides m_0 and

$$\psi : M \to D$$

is locally expressed as $\psi = (z_2)^{m_0}$ at a generic point of Θ_0, ψ naturally induces a holomorphic map

$$\psi' : M' \to D'$$

of the modified complex surface M'. (NB. M and M' are diffeomorphic.) It is easy to see that, in M', $\pi^{-1}(BDY_0)$ has w number of connected components.

Proceeding in this way on each body of S, we may assume that the number of connected components of $\pi^{-1}(BDY_\nu)$ is equal to the weight of the vertex $\eta(BDY_\nu)$ of Y, for each ν.

Number the connected components of $\pi^{-1}(BDY_\nu)$ cyclically for each BDY_ν.

Let $ARCH_0$ be any arch of S. Let BDY_0 and BDY_1 be two bodies which $ARCH_0$ connects (Possibly $BDY_0 = BDY_1$). Now a cochain $\overline{c}$ representing the cohomology class $c \in H^1_W(Y)$ indicates which connected components of $\pi^{-1}(BDY_0)$ should be joined to which ones of $\pi^{-1}(BDY_1)$ by the annuli over $ARCH_0$. (Cf. Lemma 6.5). We can realize this joining by modifying M and ψ as follows: Let l be a simple arc on $ARCH_0$ joining the two end-circles of $ARCH_0$, and let C_1 be a simple closed curve on an irreducible component of $ARCH_0$ (or on an attaching bank) which cuts l transversely once. Then we do the same process as before of cutting open M along $C_1 \times \mathbf{C}$ and re-glue the two copies $(C_1 \times \mathbf{C})_0$ and $(C_1 \times \mathbf{C})_1$ through appropiate rotation of the "fibers" $\mathbf{C}$.

Proceeding in this way on each arch in S, we will get the modified complex surface, again denoted by M, and a proper, surjective, and holomorphic map $\psi : M \to D$ having the desired aspect of the joints among the connected components of $\pi^{-1}(ARCH_\mu)$ and $\pi^{-1}(ARCH_\nu)$ indicated by the cochain $\overline{c}$.

The general fiber F_δ is now *connected* because of Condition (4). (See Lemma 6.6). The fiber F_δ has genus g because of Condition (5).

Let $f : F_\delta \to F_\delta$ be the monodromy homeomorphism associated to the degenerating family (M, D, ψ). Then f is a pseudo-periodic map of negative twist with the desired properties.

This completes the proof of Theorem 9.2. □

9.4 Concluding Remark

Let (S, Y, c) be a triple consisting of a compact, connected, and numerical chorizo space S satisfying the minimality condition; a weighted graph Y (with a collapsing map $\eta : S \to Y$); and a weighted cohomology class $c \in H^1_W(Y)$. Two such triples (S_1, Y_1, c_1) and (S_2, Y_2, c_2) are *equivalent* if there exist a numerical homeomorphism $H : S_1 \to S_2$ and a weighted isomorphism $\Psi : Y_1 \to Y_2$ such that the diagram

$$\begin{array}{ccc} S_1 & \xrightarrow{\eta_1} & Y_1 \\ {\scriptstyle H}\downarrow & & \downarrow{\scriptstyle \Psi} \\ S_2 & \xrightarrow[\eta_2]{} & Y_2 \end{array}$$

commutes and such that $\Psi^*(c_2) = c_1$.

Let $\mathscr{T}_g$ denote the set of all equivalence classes of such triples (S, Y, c) that satisfy Conditions (1)$\sim$(5) stated before Theorem 9.2.

By sending each pseudo-periodic map $f : \Sigma_g \to \Sigma_g$ $(g \geq 2)$ of negative twist to the triple $(S[f], Y[f], c[f])$, defined in Chaps. 4 and 6, we obtain a map

$$\tau : \mathscr{P}_g^- \to \mathscr{T}_g.$$

Theorems 4.2 and 6.3 ensure well-definedness and injectivity of this map, and Theorem 9.2 ensures its surjectivity (thus bijectivity).

Therefore, we have proved that every map in the following diagram is bijective (if $g \geq 2$):

$$\begin{array}{ccccc} \hat{\mathscr{S}}_g & & \overrightarrow{\tau_{\hat\rho}} & & \mathscr{T}_g \\ & \widehat{\rho} \searrow & & & \| \\ \downarrow^{\beta} & & \mathscr{P}_g^- & \overrightarrow{\tau} & \mathscr{T}_g \\ & \rho \nearrow & & & \| \\ \mathscr{S}_g & & \overrightarrow{\tau_\rho} & & \mathscr{T}_g \end{array}$$

Map β: Theorem 8.1 and Proposition 9.1.
Map $\widehat{\rho}$ (with inverse $\widehat{\sigma}$): Lemmas 9.2 and 9.3.
Map ρ: Theorem 7.2.
Map τ: Theorems 4.2, 6.3 and 9.2.

Appendix A
Periodic Maps Which Are Homotopic

The purpose of this appendix is to give a proof of Theorem 2.2. We will state the theorem again.

Theorem A.1. *Let f and f' be (orientation-preserving) periodic maps of a compact surface Σ each component of which has negative Euler characteristic. Suppose f and $f' : (\Sigma, \partial\Sigma) \to (\Sigma, \partial\Sigma)$ are homotopic as maps of pairs. Then there exists a homeomorphism $h : \Sigma \to \Sigma$ isotopic to the identity, such that $f = h^{-1} f' h$.*

Proof. First we consider the case when Σ is *connected*. Note that the quotient space $M = \Sigma/f$ is an orbifold with negative orbifold-Euler characteristic, (cf. [62]).

Case 1. The underlying space $|M|$ of M is not S^2 nor D^2.

Let $DM = M \cup \overline{M}$ be the double of M. (If $\partial M = \emptyset$, set $DM = M$.) We put a hyperbolic metric on DM so that DM admits a decomposition by a finite number of simple closed geodesics

$$G_1,\ G_2, \ldots,\ G_l$$

which satisfy the following conditions:

1. No G_i passes through a cone point (of course, this can never happen, because the cone angle is less than 2π),
2. the boundary curves of M are closed geodesics, and are members of

$$\{G_1,\ G_2, \ldots,\ G_l\},$$

3. the intersections of

$$G_1,\ G_2, \ldots,\ G_l$$

are only double points,

4. each component of $DM - \bigcup_{i=1}^{l} G_i$ is an open cell whose closure is a polygon with more than 3 edges, and

Y. Matsumoto and J.M. Montesinos-Amilibia, *Pseudo-periodic Maps and Degeneration of Riemann Surfaces*, Lecture Notes in Mathematics 2030,
DOI 10.1007/978-3-642-22534-5, © Springer-Verlag Berlin Heidelberg 2011

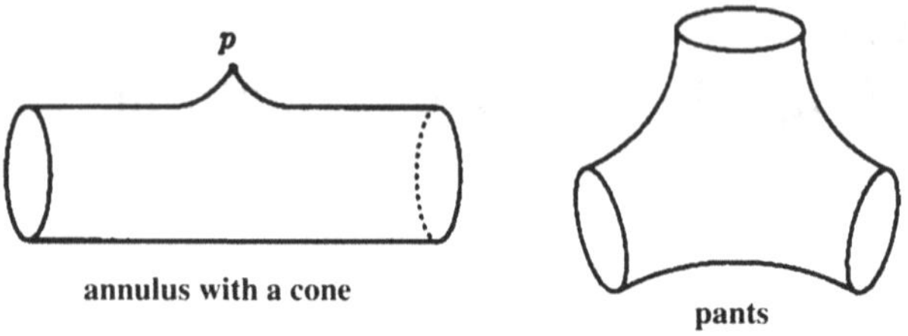

Fig. A.1 Hyperbolic parts ("annulus with a cone" and "pants")

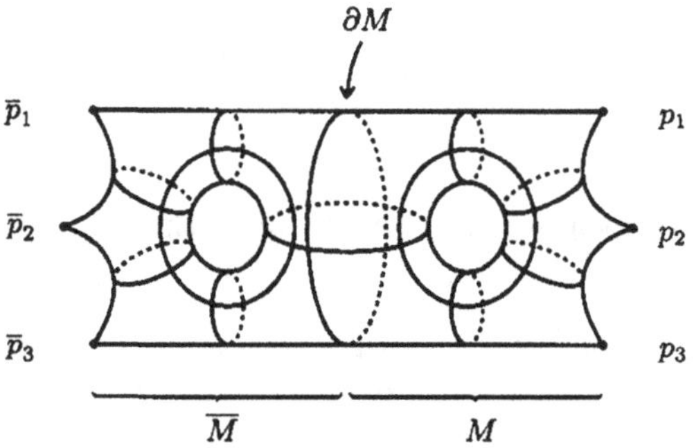

Fig. A.2 Decomposing the double DM by simple closed geodesics

5. each component of $DM - \bigcup_{i=1}^{l} G_i$ contains at most one cone point.

   Such a decomposition is certainly possible. In fact, one can start with two kinds of hyperbolic parts; annuli with one cone point and pants, both having geodesic boundaries of length 1. See Fig. A.1. Glue them along the boundaries and construct a hyperbolic orbifold homeomorphic to DM. Add further simple closed geodesics to get a desired decomposition. See Fig. A.2.

Let $D\Sigma = \Sigma \cup \overline{\Sigma}$ denote the double of Σ. The periodic map $f : \Sigma \to \Sigma$ symmetrically extends to a periodic map $D\Sigma \to D\Sigma$, which will be denoted by f again. Lift the metric on DM just constructed, to $D\Sigma$ to make the latter a (smooth) hyperbolic surface. The periodic map $f : D\Sigma \to D\Sigma$ preserves this metric σ.

Let

$$\Gamma_1,\ \Gamma_2, \ldots,\ \Gamma_m$$

be the lifts to $D\Sigma$ of the simple closed geodesics

$$G_1,\ G_2, \ldots,\ G_l.$$

The total number m of the Γ_i's might be different from the total number l of the G_i's. The simple closed geodesics

$$\Gamma_1,\ \Gamma_2, \ldots,\ \Gamma_m$$

satisfy the following conditions:

(i) No Γ_i passes a multiple point of f,
(ii) the boundary curves of Σ are members of

$$\{\Gamma_1, \Gamma_2, \ldots, \Gamma_m\},$$

(iii) the intersections of

$$\Gamma_1, \Gamma_2, \ldots, \Gamma_m$$

are only double points,
(iv) each component of $D\Sigma - \bigcup_{i=i}^{m} \Gamma_i$ is an open cell whose closure is a polygon with more than 3 edges, and
(v) each component of $D\Sigma - \bigcup_{i=1}^{m} \Gamma_i$ contains at most one cone point.
Moreover, we have the following:
(vi) No pair Γ_i, Γ_j $(i \neq j)$ are freely homotopic (because they are distinct closed geodesics. Cf. [16].)
(vii) Γ_i and Γ_j $(i \neq j)$ have minimal intersection. Cf. [16].
(viii) f preserves the configuration $\Gamma_1 \cup \Gamma_2 \cup \cdots \cup \Gamma_m$;

$$f(\bigcup_{i=1}^{m} \Gamma_i) = \bigcup_{i=1}^{m} \Gamma_i$$

due to the construction of

$$\Gamma_1, \Gamma_2, \ldots, \Gamma_m.$$

So far we have only considered the periodic map f. Now we consider the other

$$f' : \Sigma \to \Sigma.$$

Extend f' symmetrically to the double $D\Sigma$ and denote the resulting periodic map by $f' : D\Sigma \to D\Sigma$ also. Put a hyperbolic metric σ' on $D\Sigma$ which is invariant under f' and such that the boundary curves of Σ are closed geodesics.

The simple closed curves

$$\Gamma_1, \Gamma_2, \ldots, \Gamma_m$$

are no longer geodesics with respect to σ', in general. But they are freely homotopic to simple closed geodesics

$$\Gamma_1', \Gamma_2', \ldots, \Gamma_m'$$

with respect to σ'. [16, Lemma 2.3.] These curves are distinct thanks to property (vi) of the Γ_i's.

We will construct an isotopy

$$g_\tau : D\Sigma \to D\Sigma, \quad 0 \leq \tau \leq 1,$$

such that

(a) $g_0 = id_{D\Sigma}$,
(b) if $\Gamma_i = \Gamma_i'$ (for instance, a boundary curve of Σ), then $g_\tau(\Gamma_i) = \Gamma_i'$, and
(c) $g_1(\Gamma_i) = \Gamma_i'$, $i = 1, 2, \ldots, m$.

The construction is only a mimic of Lemmmas 2.4, 2.5 of Casson's lecture notes, [16], where the case $m = 2$ is treated, and is done essentially by an innermost arc argument. We proceed by induction. Suppose $\Gamma_i = \Gamma_i'$ $(i = 1, 2, \ldots, k)$ for some $k < m$. We will find an isotopy which starts with the identity, setwise preserves $\Gamma_i (= \Gamma_i')$ for $i = 1, 2, \ldots, k$, and at the final stage sends Γ_{k+1} to Γ_{k+1}'. (It will be helpful to consider

$$\Gamma_1',\ \Gamma_2', \ldots,\ \Gamma_m'$$

as patterns drawn on a floor, that remain fixed, and

$$\Gamma_1,\ \Gamma_2, \ldots,\ \Gamma_m$$

as loops laid on it, that can move.) First we isotop Γ_{k+1} to separate it from Γ_{k+1}', and then using the annulus between Γ_{k+1} and Γ_{k+1}' move Γ_{k+1} onto Γ_{k+1}', as we will sketch now. A typical move is performed through the shaded disk in Fig. A.3.

The geodesics

$$\Gamma_1', \ldots,\ \Gamma_k'\ (\Gamma_1, \ldots,\ \Gamma_k)$$

have minimal intersection with the geodesic Γ_{k+1}' (cf. [16, Lemma 2.5]). Also they have minimal intersection with Γ_{k+1} because of property (vii) of Γ_i's. Thus as in Casson's Lemma 2.5, $\Gamma_i' \cap$ (the shaded disk) is a family of arcs passing "through" the disk from "top to bottom", for $i = 1, 2, \ldots, k$. But if there were a situation as shown in Fig. A.4, we would have an obstacle; we could not move Γ_{k+1} "along" $\Gamma_1', \ldots, \Gamma_k'$.

This "bad" situation is, however, prohibited by property (iv) of the Γ_i's. Therefore, we have the situation of Fig. A.3, and can move Γ_{k+1} through the shaded disk "along"

$$\Gamma_1',\ \Gamma_2', \ldots,\ \Gamma_k'\ (= \Gamma_1,\ \Gamma_2, \ldots,\ \Gamma_k).$$

This completes the inductive step, and we have obtained an isotopy

$$g_\tau : D\Sigma \to D\Sigma, \quad 0 \leq \tau \leq 1$$

satisfying conditions (a), (b), (c) stated above.

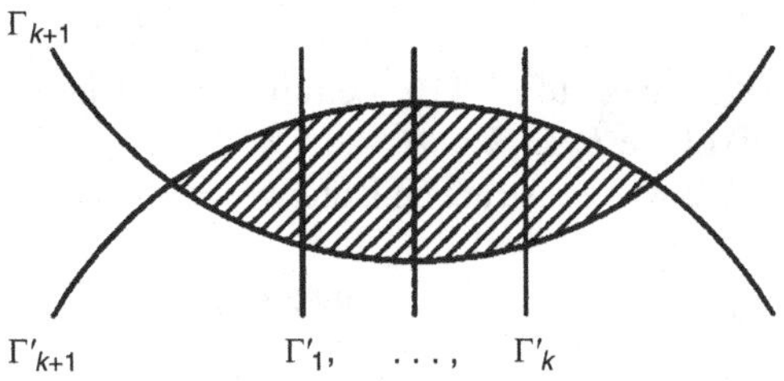

Fig. A.3 A move of Γ_{k+1}

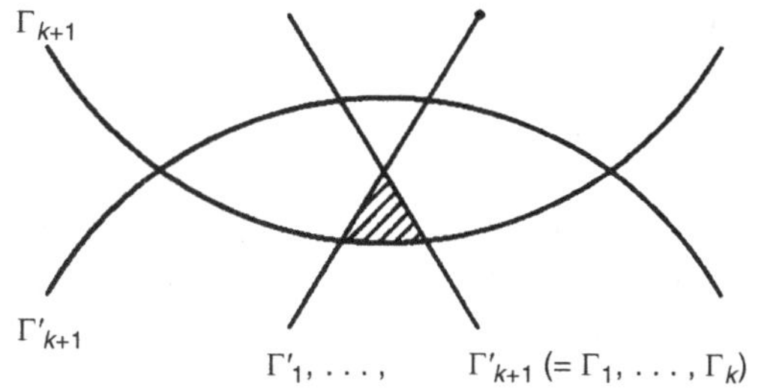

Fig. A.4 A situation in which the move of Γ_{k+1} is obstructed

Let $g : D\Sigma \to D\Sigma$ be the final stage g_1 of the isotopy. Then

$$g(\Gamma_1' \cup \Gamma_2 \cup \cdots \cup \Gamma_m) = \Gamma_1' \cup \Gamma_2' \cup \cdots \cup \Gamma_m'.$$

Lemma A.1. $f'(\Gamma_1' \cup \Gamma_2' \cup \cdots \cup \Gamma_m') = \Gamma_1' \cup \Gamma_2' \cup \cdots \cup \Gamma_m'$.

Proof. First note that

$$\begin{aligned} f'(\Gamma_i') &\simeq f'(\Gamma_i) && (\text{because} \quad \Gamma_i' \simeq \Gamma_i) \\ &\simeq f(\Gamma_i) && (\text{because} \quad f' \simeq f) \\ &= \Gamma_j && (\text{for some } j, \text{ because of property (viii) of the } \Gamma_i\text{'s}) \\ &\simeq \Gamma_j' && (\text{because } \Gamma_j \simeq \Gamma_j'), \end{aligned}$$

where "$\simeq$" denotes "is freely homotopic to".

The metric σ' on $D\Sigma$ is invariant under f', so $f'(\Gamma_i')$ is a simple closed geodesic as well as Γ_i's. In a free homotopy class of simple closed curves, there is only one simple closed geodesic, (cf. [16, Lemma 2.4]). Therefore, $f'(\Gamma_i') = \Gamma_j'$. □

Let $\mathbf{H}^2$ be the hyperbolic plane which is the universal covering of $(D\Sigma, \sigma')$. Let $D = \mathbf{H} \cup S_\infty$ be its compactification to the unit disk. Let

$$\tilde{g}_\tau : \mathbf{H}^2 \to \mathbf{H}^2, \quad 0 \le \tau \le 1$$

be the lifted isotopy of g_τ starting from $\tilde{g}_0 = id_{\mathbf{H}^2}$. Then

$$\tilde{g} := \tilde{g}_1 : \mathbf{H}^2 \to \mathbf{H}^2$$

is a lift of

$$g : D\Sigma \to D\Sigma.$$

By Nielsen [52], the isotopy

$$\tilde{g}_\tau : \mathbf{H}^2 \to \mathbf{H}^2$$

extends to an isotopy

$$(\tilde{g}_\tau)^\wedge : D \to D,$$

and being a lift of an isotopy of a *compact* surface $D\Sigma$, the restriction of $(\tilde{g}_\tau)^\wedge$ to S_∞ is *constant* because $\tilde{g}_\tau$ varies equivariantly with respect to the group of superpositions of the covering $\mathbf{H}^2 \to D\Sigma$, and the variation is determined by the restriction of $\tilde{g}_\tau$ to a compact set (because $D\Sigma$ is compact). Since the euclidean size of a fundamental domain for $D\Sigma$ tends to zero if it goes to infinity, then $\tilde{g}_\tau$ at S_∞ does not depend on τ. Thus $(\tilde{g}_\tau)^\wedge|S_\infty = id|S_\infty$.

Let

$$\tilde{f} : \mathbf{H}^2 \to \mathbf{H}^2$$

be a lift of

$$f : D\Sigma \to D\Sigma,$$

and

$$(\tilde{f})^\wedge : D \to D$$

its extention. Since

$$f \simeq f' : D\Sigma \to D\Sigma,$$

$\tilde{f}$ is homotopic, through a lifted homotopy, to a lift $\tilde{f}'$ of f'. For the same reason as above, we have $(\tilde{f})^\wedge|S_\infty = (\tilde{f}')^\wedge|S_\infty$.

Now let us regard

$$\Gamma_1 \cup \Gamma_2 \cup \cdots \cup \Gamma_m$$

and

$$\Gamma_1' \cup \Gamma_2' \cup \cdots \cup \Gamma_m'$$

as finite graphs Γ and Γ', respectively, drawn on $D\Sigma$. They decompose $D\Sigma$ into finite cell complexes. Consider their lifts $\tilde{\Gamma}$, $\tilde{\Gamma}'$ in $\mathbf{H}^2$. $\tilde{\Gamma}$ consists of lifts of the Γ_i's, and $\tilde{\Gamma}'$ of lifts of the Γ_i's. (Each lift of Γ_i' is a geodesic line.) $\tilde{\Gamma}$ and $\tilde{\Gamma}'$ decompose $\mathbf{H}^2$ into locally finite cell complexes. $\tilde{g} : \mathbf{H}^2 \to \mathbf{H}^2$ preserves these patterns: $\tilde{g}(\tilde{\Gamma}) = \tilde{\Gamma}'$.

Let $\tilde{\Gamma}^{(0)}$ and $\tilde{\Gamma}'^{(0)}$ be the set of vertices of $\tilde{\Gamma}$ and $\tilde{\Gamma}'$, respectively.

Lemma A.2. *$\tilde{g} \mid \tilde{\Gamma}^{(0)} : \tilde{\Gamma}^{(0)} \to \tilde{\Gamma}'^{(0)}$ is equivariant with respect to $\tilde{f} \mid \tilde{\Gamma}^{(0)}$ and $\tilde{f}' \mid \tilde{\Gamma}'^{(0)}$.*

Proof. Take a vertex $\tilde{v} \in \tilde{\Gamma}^{(0)}$. $\tilde{v}$ is an intersection point of two infinite curves $\tilde{\Gamma}_i$ and $\tilde{\Gamma}_j$, which are lifts of Γ_i and Γ_j. By property (iii) of the Γ_i's, the pair $\{\tilde{\Gamma}_i, \tilde{\Gamma}_j\}$ is uniquely determined by $\tilde{v}$. We have

$$\tilde{g}\,\tilde{f}(\tilde{\Gamma}_i) = \tilde{f}'\,\tilde{g}(\tilde{\Gamma}_i) \quad \text{and} \quad \tilde{g}\,\tilde{f}(\tilde{\Gamma}_j) = \tilde{f}'\,\tilde{g}(\tilde{\Gamma}_j).$$

(*Proof.* Let $\inf(\tilde{\Gamma}_i)$ denote the pair of the two "infinite" points of $\tilde{\Gamma}_i$ in S_∞. Then

$$\inf(\tilde{g}\,\tilde{f}(\tilde{\Gamma}_i)) = (\tilde{g})^\wedge(\tilde{f})^\wedge(\inf(\tilde{\Gamma}_i)) = (\tilde{f}')^\wedge(\tilde{g})^\wedge(\inf(\tilde{\Gamma}_i)) = \inf(\tilde{f}'\,\tilde{g}(\tilde{\Gamma}_i))$$

because $(\tilde{g})\hat{}\,|\,S_\infty = id$ and

$$(\tilde{f})\hat{}\,|\,S_\infty = (\tilde{f}')\hat{}\,|\,S_\infty.$$

$\tilde{g}\,\tilde{f}(\tilde{\Gamma}_i)$ and $\tilde{f}'\,\tilde{g}(\tilde{\Gamma}_i)$ are geodesic lines in $\mathbf{H}^2$ with the same pair of infinite points. Then they coincide. Similarly, $\tilde{g}\,\tilde{f}(\tilde{\Gamma}_j) = \tilde{f}'\,\tilde{g}(\tilde{\Gamma}_j)$.)

Therefore,

$$\{\tilde{g}\,\tilde{f}(\tilde{v})\} = \tilde{g}\,\tilde{f}(\tilde{\Gamma}_i) \cap \tilde{g}\,\tilde{f}(\tilde{\Gamma}_j) = \tilde{f}'\,\tilde{g}(\tilde{\Gamma}_i) \cap \tilde{f}'\,\tilde{g}(\tilde{\Gamma}_j) = \{\tilde{f}'\,\tilde{g}(\tilde{v})\},$$

which proves $\tilde{g}\,\tilde{f}(\tilde{v}) = \tilde{f}'\,\tilde{g}(\tilde{v})$. □

Remember that the $\tilde{\Gamma}_i$'s are geodesic lines with respect to the lifted metric $\tilde{\sigma}$ of σ. Using Lemma A.2, we can find an isotopy

$$\tilde{g}^{(1)}_\tau : \mathbf{H}^2 \to \mathbf{H}^2, \quad 0 \le \tau \le 1,$$

such that $\tilde{g}^{(1)}_0 = \tilde{g}$, $\tilde{g}^{(1)}_\tau(\tilde{\Gamma}) = \tilde{\Gamma}'$, $\tilde{g}^{(1)}_\tau$ is equivariant with respect to the group of covering translations of $\mathbf{H}^2 \to D\Sigma$, and the final stage $\tilde{g}^{(1)}_1$ is "linear" from each edge of $\tilde{\Gamma}$ to an edge of $\tilde{\Gamma}'$ with respect to $\tilde{\sigma}$ and $\tilde{\sigma}'$ (the lifted metric of σ'). Then

$$\tilde{g}^{(1)}_1|\tilde{\Gamma} : \tilde{\Gamma} \to \tilde{\Gamma}'$$

is equivariant with respect to $\tilde{f}|\tilde{\Gamma}$ and $\tilde{f}'|\tilde{\Gamma}'$.

Finally by the Alexander trick, we can deform $\tilde{g}^{(1)}_1$ within each cell and obtain an isotopy

$$\tilde{g}^{(2)}_\tau : \mathbf{H}^2 \to \mathbf{H}^2, \quad 0 \le \tau \le 1,$$

such that $\tilde{g}^{(2)}_0 = \tilde{g}^{(1)}_1$, $\tilde{g}^{(2)}_\tau|\tilde{\Gamma} = \tilde{g}^{(1)}_\tau|\tilde{\Gamma}$, $\tilde{g}^{(2)}_\tau$ is equivariant with respect to the group of covering translations of

$$\mathbf{H}^2 \to D\Sigma,$$

and the final stage

$$\tilde{g}^{(2)}_1 : \mathbf{H}^2 \to \mathbf{H}^2$$

is equivariant with respect to $\tilde{f}$ and $\tilde{f}'$. This homeomorphism $\tilde{g}^{(2)}_1$ projects down to a homeomorphism

$$h : D\Sigma \to D\Sigma.$$

The restricted homeomorphism

$$h|\Sigma : \Sigma \to \Sigma$$

is the one whose existence is asserted by Theorem 2.2. This completes the proof of Case 1.

Case 2. The underlying space $|M|$ of M is S^2.

The proof will be accompanied by several lemmas.

Lemma A.3. *The orders of f and $f' : \Sigma \to \Sigma$ are equal.*

Proof. Let $L(g)$ denote the Lefschetz number of a homeomorphism $g : \Sigma \to \Sigma$. Then the order of f is equal to the smallest positive integer n such that $L(f^n) < 0$ because if $f^n \neq id_\Sigma$, $L(f^n)$ is equal to the number of the fixed points of f^n (see [23, pp. 130, 121]) which is non-negative, while $L(id_\Sigma) = \chi(\Sigma) < 0$. Thus our assumption $f \simeq f'$ implies that their orders coincide. □

Lemma A.4. *There is a bijective correspondence between the set of multiple points of f and the same set of f' which preserves the valencies.*

Proof. We impose a hyperbolic metric on Σ and identify the universal covering $\tilde{\Sigma}$ with $\mathbf{H}^2$, which is compactified to $D = \mathbf{H}^2 \cup S_\infty$ as before. Let $F(f)$ denote the set of fixed points of f. Take a point $p_0 \in F(f)$ and lift it to $\tilde{p}_0 \in \mathbf{H}^2$. Let

$$\tilde{f} : \mathbf{H}^2 \to \mathbf{H}^2$$

be a lift of f which fixes $\tilde{p}_0$. By Nielsen [50, Sect. 2], $\tilde{f}$ extends to a homeomorphism

$$(\tilde{f})\hat{} : D \to D.$$

The lift

$$\tilde{f} : \mathbf{H}^2 \to \mathbf{H}^2$$

is a periodic map because it is a lift of a periodic map and it has a fixed point. Then

$$(\tilde{f})\hat{} : D \to D$$

is also a periodic map. By [31], an orientation-preserving periodic map of a 2-disk D is conjugate to a rotation. In particular, $(\tilde{f})\hat{}|S_\infty$ has no fixed points.

By our assumption

$$f \simeq f' : \Sigma \to \Sigma,$$

$\tilde{f}$ is homotopic, through a lifted homotopy, to a lift $\tilde{f}'$ of f'. Then by the same argument as in Case 1, $(\tilde{f})\hat{}|S_\infty = (\tilde{f}')\hat{}|S_\infty$. By Brouwer's fixed point theorem,

$$(\tilde{f}')\hat{} : D \to D$$

has a fixed point. But $(\tilde{f}')|_{S_\infty}$ $(= (\tilde{f})\hat{}|_{S_\infty})$ has no fixed points, so the fixed point of $\tilde{f}'$ is in $\mathbf{H}^2$. Then the same argument as for $\tilde{f}$ can apply to $\tilde{f}'$, and

$$(\tilde{f}')\hat{} : D \to D$$

is conjugate to a rotation. Thus the fixed point $\tilde{p}_0'$ of $\tilde{f}'$ is uniquely determined in $\mathbf{H}^2$, which projects to a fixed point p_0' of f'. The correspondence

$$\varphi : F(f) \to F(f')$$

is defined by sending p_0 to p'_0.

φ is independent of the choice of the lift $\tilde{p}_0$, but it might depend on the homotopy between f and f'. We will fix the homotopy throughout the argument. Clearly φ is bijective because the roles of f and f' are symmetric.

The valency of p_0 (resp. p'_0) with respect to f (resp. f') is the same as the valency of $\tilde{p}_0$ (resp. $\tilde{p}'_0$) with respect to $\tilde{f}$ (resp. $\tilde{f}'$), which can be read off from the action of $(\tilde{f})\hat{}$ (resp. $(\tilde{f}')\hat{}$) on S_∞. But $(\tilde{f})\hat{}|S_\infty = (\tilde{f}')\hat{}|S_\infty$. Thus the valency of p_0 is equal to the valency of $p'_0 = \varphi(p_0)$.

Similarly for each factor m of the order of f we can construct a bijective correspondence between $F(f^m)$ and $F((f')^m)$. This correspondence preserves the valencies exactly as above. This completes the proof of Lemma A.4. □

Lemmmas A.3, A.4 and Nielsen's theorem (Theorem 1.2) imply that $f : \Sigma \to \Sigma$ and $f' : \Sigma \to \Sigma$ are conjugate. (Remember that we are considering a closed Σ in Case 2.) In particular, we have

Lemma A.5. *$M = \Sigma/f$ and $M' = \Sigma'/f'$ are homeomorphic as orbifolds.*

Let

$$p_1, \ p_2, \dots, \ p_s, \quad s \geq 3,$$

be the cone points of M with valencies

$$(m_1, \lambda_1, \sigma_1), \ (m_2, \lambda_2, \sigma_2), \dots, \ (m_s, \lambda_s, \sigma_s),$$

respectively. Consider a polygon P (s-gon) in $\mathbf{H}^2$ whose angles are

$$\pi/\lambda_1, \ \pi/\lambda_2, \dots, \ \pi/\lambda_s.$$

See Fig. A.5. (Such a P exists because $\chi^{\text{orb}}(M) < 0$.) M can be considered as a hyperbolic "bi-hedron" having two faces, each congruent with P.

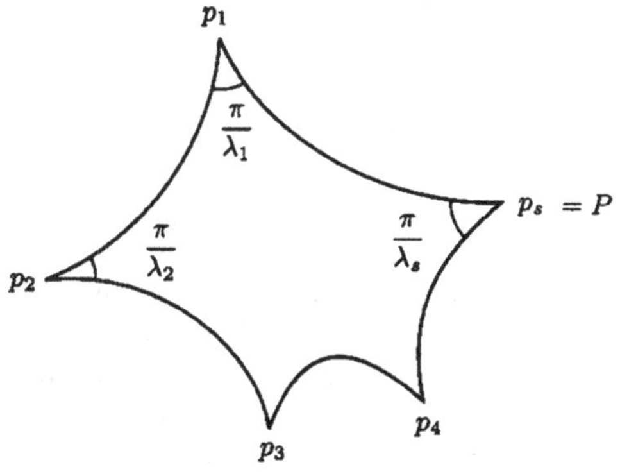

Fig. A.5 A hyperbolic polygon

Let σ be the hyperbolic metric on Σ obtained by lifting the bi-hedron metric on M through the projection $\Sigma \to M$. We lift this metric σ farther to the universal covering $\tilde{\Sigma}$ to make it a hyperbolic plane $\mathbf{H}^2(\sigma)$. The plane $\mathbf{H}^2(\sigma)$ is tessellated by tiles, each congruent with P. The fundamental region for the action of $\pi_1^{\mathrm{orb}}(M)$ ([62, Sect. 13]) is $F = P \cup \overline{P}$, where $\overline{P}$ is a flipped P having an edge in common with it.

Let r_i be the rotation (in $\mathbf{H}^2(\sigma)$) of angle $2\pi/\lambda_i$ centered at vertex $p_i (\in P)$, $i = 1, 2, \ldots, s$. It is known that the rotations r_1, r_2, ..., r_s generate the group $\pi_1^{\mathrm{orb}}(M)$, the orientation-preserving automorphism group of the tessellation, (cf. Milnor [45] or [41]). Moreover, from our construction, r_i is a lift of

$$f^{\sigma_i m_i} : \Sigma \to \Sigma, \quad i = 1,\ 2, \ldots,\ s.$$

Let

$$f_\tau : \Sigma \to \Sigma, \quad 0 \leq \tau \leq 1,$$

be a homotopy between f and f' : $f_0 = f$, $f_1 = f'$. Then, through a lifted homotpy

$$(f_\tau^{\sigma_i m_i})^{\tilde{}} : \mathbf{H}^2(\sigma) \to \mathbf{H}^2(\sigma),$$

r_i is homotopic to a homeomorphism

$$r_i' : \mathbf{H}^2(\sigma) \to \mathbf{H}^2(\sigma).$$

This r_i' is a lift of $(f')^{\sigma_i m_i}$, satisfies $(r_i')^{\hat{}}|S_\infty = (r_i)^{\hat{}}|S_\infty$, and is topologically equivalent to a rotation. (See Proof of Lemma A.4.)

Let p_i' be the center of the "rotation" r_i'. We denote the cone point of M' to which p_i' projects by the same notation p_i'. Thus we have obtained the following correspondence:

$$M \ni p_i \longleftrightarrow p_i' \in M', \quad i = 1, 2, \ldots, s.$$

By Lemma A.4, this correspondence preserves the valencies. We impose a structure of a hyperbolic bi-hedron on M' whose faces P', $\overline{P}'$ are congruent with P, $\overline{P}$ preserving the above correspondence of the vertices. Let σ' be the hyperbolic metric on Σ obtained by lifting the bi-hedron metric on M' through the projection $\Sigma \to M'$. We lift σ' to $\tilde{\Sigma}$ to make it a hyperbolic plane $\mathbf{H}^2(\sigma')$, in which the polygon P' is inscribed with vertices

$$p_1',\ p_2', \ldots,\ p_s'.$$

In $\mathbf{H}^2(\sigma)$, the topological rotation r_i' is a genuine rotation with center p_i' of angle $2\pi/\lambda_i$, $i = 1, 2, \ldots, s$. The plane $\mathbf{H}^2(\sigma')$ is tessellated by tiles, each congruent with P'. The rotations r_1', r_2', ..., r_s' generate the group $\pi_1^{\mathrm{orb}}(M')$, the orientation-preserving automorphism group of the tessellation. Since P' is congruent with P, the group $\pi_1^{\mathrm{orb}}(M')$ is isomorphic to $\pi_1^{\mathrm{orb}}(M)$ via the correspondence

$$r_i \longleftrightarrow r_i', \quad i = 1, 2, \ldots, s.$$

Note that $\mathbf{H}^2(\sigma)$ and $\mathbf{H}^2(\sigma')$ are merely different "pictures" on the same space $\tilde{\Sigma}$, so $\pi_1^{\text{orb}}(M)$ and $\pi_1^{\text{orb}}(M')$ are considered as subgroups of $Homeo(\tilde{\Sigma})$, the group of all self-homeomorphisms of $\tilde{\Sigma}$. If we fix the action of $\pi_1(\Sigma)$ on $\tilde{\Sigma}$, then $\pi_1(\Sigma)$ is also a subgroup of $Homeo(\tilde{\Sigma})$, contained in $\pi_1^{\text{orb}}(M) \cap \pi_1^{\text{orb}}(M')$. Moreover, the action of $\pi_1(\Sigma)$ is isometric, with respect to $\mathbf{H}^2(\sigma)$ *and* at the same time with respect to $\mathbf{H}^2(\sigma')$.

Lemma A.6. *The isomorphism between $\pi_1^{\text{orb}}(M)$ and $\pi_1^{\text{orb}}(M')$ given by the correspondence $r_i \longleftrightarrow r_i'$ $(i = 1, 2, \ldots, s)$ restricts to the identity on $\pi_1(\Sigma)$.*

Proof. Take an element $g \in \pi_1(\Sigma)$. Since $\pi_1(\Sigma) < \pi_1^{\text{orb}}(M')$, g can be written as a product of $r_1', r_2', \ldots, r_s'$:

$$g = \psi(r_1', r_2', \ldots, r_s').$$

We will show that

$$g = \psi(r_1, r_2, \ldots, r_s).$$

For this, compactify $\mathbf{H}^2(\sigma)$ to $D = \mathbf{H}^2(\sigma) \cup S_\infty$ as before. Then

$$\hat{g}|S_\infty = \psi(r_1', r_2', \ldots, r_s')\hat{}\,|S_\infty = \psi(r_1, r_2, \ldots, r_s)\hat{}\,|S_\infty.$$

Since both g and $\psi(r_1, r_2, \ldots, r_s)$ are isometries of $\mathbf{H}^2(\sigma)$, we have

$$g = \psi(r_1, r_2, \ldots, r_s).$$

□

Let us construct a homeomorphism $\tilde{h} : \tilde{\Sigma} \to \tilde{\Sigma}$ which is equivariant with respect to the actions of $\pi_1^{\text{orb}}(M)$ and $\pi_1^{\text{orb}}(M')$. The construction is obvious. First, map the fundamental region $F = P \cup \overline{P}$ "isometrically" to the fundamental region $F' = P \cup \overline{P}'$. Then extend it equivariantly to the whole space.

By Lemma A.6, we have $\tilde{h}\, g = g\, \tilde{h}$ for all $g \in \pi_1(\Sigma)$. Thus $\tilde{h}$ projects to a homeomorphism $h : \Sigma \to \Sigma$.

Lemma A.7. *$h : \Sigma \to \Sigma$ satisfies $f = h^{-1} f' h$.*

Proof. Let $\tilde{f} : \tilde{\Sigma} \to \Sigma$ be a lift of $f : \Sigma \to \Sigma$. Since $\tilde{f}$ preserves the tessellation, $\tilde{f}$ is written as a product of $r_1, r_2, \ldots, r_s$:

$$\tilde{f} = \varphi(r_1, r_2, \ldots, r_s).$$

Then

$$\varphi(r_1', r_2', \ldots, r_s')$$

is a lift of f'; denote it by $\tilde{f}'$. Since $\tilde{h}$ is equivariant with respect to the actions of $\pi_1^{\mathrm{orb}}(M)$ and $\pi_1^{\mathrm{orb}}(M')$, we have

$$\tilde{h}\,\tilde{f} = \tilde{h}\,\varphi(r_1, r_2, \ldots, r_s) = \varphi(r_1', r_2', \ldots, r_s')\,\tilde{h} = \tilde{f}'\,\tilde{h},$$

so $h\,f = f'\,h$ as asserted. □

Lemma A.8. *$h : \Sigma \to \Sigma$ is isotopic to the identity.*

Proof. It will be sufficient to prove $(\tilde{h})\hat{}|S_\infty = id.$, because then $h : \Sigma \to \Sigma$ preserves the free homotopy class of every simple closed curve.

Let g be any element of $\pi_1(\Sigma)$ different from 1. Then $g : \mathbf{H}^2 \to \mathbf{H}^2$ is a hyperbolic transformation, and for any point $x \in \mathbf{H}^2$, $g^n(x)$ converges (in D) to a definite point $V_g \in S_\infty$ as $n \to +\infty$. Also $g^n(x)$ converges to another definite point $U_g \in S_\infty$ as $n \to -\infty$. Nielsen [50, Sect. 1] calles U_g, V_g the *negative* and the *positive* fundamental points of g.

Then, in D, we have

$$(\tilde{h})\hat{}(V_g) = (\tilde{h})\hat{}(\lim_{n\to\infty} g^n(x)) = \lim_{n\to\infty} (\tilde{h})(g^n(x)) = \lim_{n\to\infty} g^n(\tilde{h})(x) = V_g.$$

Also $(\tilde{h})\hat{}(U_g) = U_g$. But the set of fundamental points

$$\{U_g \mid g \in \pi_1(\Sigma)\} \cup \{V_g \mid g \in \pi_1(\Sigma)\}$$

is *dense* is S_∞ ([53, Sect. 1 Case b)]). This proves $(\tilde{h})\hat{}|S_\infty = id.$ □

The proof is completed in Case 2.

Case 3. The underlying space $|M|$ of M is D^2.

Making the double DM, the proof is reduced to Case 2.

Now we have completed the proof in the case when Σ is *connected.*

In the general case when Σ is not necessarily connected, divide the set of the components of Σ into cycles under the permutation caused by f. Since f' is homotopic to f, f' causes the same permutation. Then in each cycle we can argue just as in the proof of Theorem 2.3(ii). (Beware that we need here the "homotopy implies isotopy" theorem, [10, 21].) This completes the proof of Theorem A.1. □

References

1. A'Campo, N.: Sur la monodromie des singularités isolées d'hypersurfaces complexes. Invent. Math. **20**, 147–169 (1973)
2. A'Campo, N.: La fonction zêta d'une monodromie. Comment. Math. Helvetici **50**, 233–248 (1975)
3. Arakawa, T., Ashikaga, T.: Local splitting families of hyperelliptic pencils I. Tohoku Math. J. **53**, 369–394 (2001): II. Nagoya Math. J. **175**, 103–124 (2004)
4. Asada, M., Matsumoto, M., Oda, T.: Local monodromy on the fundamental groups of algebraic curves along a degenerate stable curve. J. Pure Appl. Algebra. **103**, 235–283 (1995)
5. Ashikaga, T.: Local signature defect of fibered complex surfaces via monodromy and stable reduction. Comment. Math. Helv. **85**, 417–461 (2010)
6. Ashikaga, T., Konno, K.: Global and local properties of pencils of algebraic curves. In: Usui, S., et al. (eds.) Algebraic Geometry 2000 Azumino Adv. Studies in Pure Math. **36**, 1–49 (2002)
7. Ashikaga, T., Ishizaka, M.: Classification of degenerations of curves of genus three via Matsumoto-Montesinos' theorem. Tohoku Math. J. **54**, 195–226 (2002)
8. Ashikaga, T., Endo, H.: Various aspects of degenerate families of Riemann surfaces, Sugaku Expositions **19**, Amer. Math. Soc. 171–196 (2006)
9. Ashikaga, T., Ishizaka, M.: A geometric proof of the reciprocity law of Dedekind sum. Unpublished note (2009)
10. Bear, R.: Isotopie von Kurven auf orientierbaren, geschlossenen Flächen und ihr Zusammenhang mit der topologischen Deformation der Flächen. J. Reine Angew. Math. **159**, 101–111 (1928)
11. Bers, L.: Space of degenerating Riemann surfaces. Annals of Mathematics Studies, vol. 79, pp. 43–55. Princeton U.P., Princeton, New Jersey (1975)
12. Bers, L.: An extremal problem for quasiconformal mappings and a theorem of Thurston. Acta Math. **141**, 73–98 (1978)
13. Birman, J.S.: Braids, Links, and Mapping Class Groups. Annals of Mathematics Studies, vol. 82. Princeton U.P., Princeton, New Jersey (1974)
14. Brieskorn, E.: Die Monodromie der isolierten Singularitäten von Hyperflächen. Manuscritpta math. **2**, 103–161 (1970)
15. Burde, G., Zieschang, H.: Knots, de Gruyter Studies in Mathematics, vol. 5. Walter de Gruyter, Berlin, New York (1985)
16. Casson, A.J., Bleiler, S.A.: Automorphisms of Surfaces after Nielsen and Thurston. Cambridge U.P., Cambridge, New York, New Rochelle, Melbourne, Sydney (1988)
17. Clements, C.H.: Picard-Lefschetz theorem for families of nonsingular algebraic varieties acquiring ordinary singularities. Trans. Amer. Math. Soc. **136**, 93–108 (1969)
18. Deligne, P., Mumford, D.: The irreducibility of the space of curves of given genus. Publ. Math. I.H.E.S., **36** , 75–110 (1969)

Y. Matsumoto and J.M. Montesinos-Amilibia, *Pseudo-periodic Maps and Degeneration of Riemann Surfaces*, Lecture Notes in Mathematics 2030, DOI 10.1007/978-3-642-22534-5,

19. Earle, C.J., Sipe, P.L.: Families of Riemann surfaces over the punctured disk. Pacific J. Math. **150**, 79–86 (1991)
20. Eisenbud, D., Neumann, W.: Three-dimensional link theory and invariants of plane curve singularities. Annals of Mathematics Studies, vol. 110. Princeton U.P. Princeton, NJ (1985)
21. Epstein, D.B.A.: Curves on 2-manifolds and isotopies. Acta Math. **115**, 83–107 (1966)
22. Gilman, J.: On the Nielsen type and the classification for the mapping class group. Adv. Math. **40**, 68–96 (1981)
23. Guillemin, V., Pollack, A.: Differential Topology. Prentice-Hall, Inc., Englewood Cliffs, New Jersey (1974)
24. Handel, M., Thurston, W.: New Proofs of Some results of Nielsen. Adv. Math. **56**, 173–191 (1985)
25. Holzapfel, R.-P.: Chern number relations for locally abelian Galois coverings of algebraic surfaces. Math. Nachr. **138**, 263–292 (1988)
26. Imayoshi, Y.: Holomorphic families of Riemann surfaces and Teichmüller spaces. In: Riemann Surfaces and Related Topics. Annals of Mathematics Studies, vol. 97, pp. 277–300 (1981)
27. Imayoshi, Y.: A construction of holomorphic families of Riemann surfacs over the punctured disk with given monodromy. In: Papadopoulos, A. (ed.) A Handbook of Teichmüller Theory, vol. II, pp. 93–130. European Mathematical Society, Geneva (2009)
28. Imayoshi, Y.: Math. Rev. MR1217354(94h:30057)
29. Ito, T.: Splitting of singular fibers in certain holomorphic fibrations. J. Math. Sci. Univ. Tokyo **9**, 425–486 (2002)
30. Kerckhoff, S.P.: The Nielsen realization problem. Ann. Math. **117**(2), 235–265 (1983)
31. de Kerekjarto, B.: Sur les groupes compacts de transformations topologiques des surfaces. Acta Math. **74**, 129–173 (1941)
32. Kodaira, K.: On compact analytic surfaces II. Ann. Math. **77**, 563–626 (1963)
33. Laufer, H.B.: Normal two-dimensional singularities. Annals of Mathematical Studies, vol. 71. Princeton, U.P., Princeton, New Jersey (1971)
34. Lê D.T.: Sur les noeuds algébriques. Compositio Math. **25**, 281–321 (1972)
35. Lê D.T., Michel F., Weber C.: Courbes polaires et topologie des courbes planes. Ann. Sci. École Norm. Sup. **24**(4), 141–169 (1991)
36. Magnus, W., Karras, A., Solitar, D.: Combinatorial Group Theory, 2-nd edn. Dover Publication, Inc., New York (1976)
37. Matsumoto, Y.: Good torus fibrations. Contemp. Math. **35**, 375–397 (1984)
38. Matsumoto, Y.: Lefschez fibrations of genus two – A topological approach –. In: Kojima, S., et al. (eds.) Topology and Teichmüller spaces, Proceedings of the 37-th Taniguchi Symposium held in Finland, pp. 123–148. World Scientific (1996)
39. Matsumoto, Y.: Splitting of certain singular fibers of genus two. Boletín de la Soc. Mat. Mexicana, Special issue, **10**(3), 331–355 (2004)
40. Matsumoto, Y.: Topology of degeneration of Riemann surfaces. In: Brasselet, J.-P., et al. (ed.) Singularities in Geometry and Topology, Proceedings of the Trieste Singularity Summer School and Workshop, pp. 388–393. World Scientific (2007)
41. Matsumoto, Y., Montesinos-Amilibia, J.M.: A proof of Thurston's uniformization theorem of geometric orbifolds. Tokyo J. Math. **14**, 181–196 (1991)
42. Matsumoto, Y., Montesinos-Amilibia, J.M.: Pseudo-periodic homeomorphisms and degeneration of Riemann surfaces. Bull. Am. Math. Soc. (N.S.) **30**, 70–75 (1994)
43. Michel, F., Weber, C.: On the monodromies of a polynomial map from $\mathbf{C}^2$ to $\mathbf{C}$. Topology **40**, 1217–1240 (2001)
44. Milnor, J.: Singular points of complex hypersurfaces. Ann. Math. Studies **61** (1968)
45. Milnor, J.: On the 3-dimensional Brieskorn manifolds $M(p,q,r)$. In: Neuwirth, L.P. (ed.) Knots, Groups, and 3-Manifolds, Annals of Mathematical Studies, vol. 84, pp. 175–225. Princeton, U.P., Princeton, New Jersey (1975), Papers Dedicated to the Memory of R.H. Fox
46. Montesinos-Amilibia, J.M.: Lectures on 3-fold simple coverings and 3-manifolds. Contemp. Math., **44**, 157–177 (1985)
47. Myerson, G.: On semi-regular finite continued fractions. Arch. Math. **48**, 420–425 (1987)

48. Namikawa, Y.: Studies on degeneration. In: Classification of Algebraic Varieties and Compact Complex Manifolds, Lecture Notes in Mathematics, vol. 412. Springer-Verlag, Berlin, Heidelberg, New York (1974)
49. Namikawa, Y., Ueno, K.: The complete classification of curves in pencils of curves of genus two. Manuscripta Math. **9**, 143–186 (1973)
50. Nielsen, J.: Investigations in the topology of closed orientable surfaces, I. Translation by John Stillwell of: Untersuchungen zur Topologie der geschlossenen zweiseitigen Flächen, [Acta Math., **50**, 189–358 (1927)], [Collected Math. Papers, **1**, Birkhäuser, Boston, Basel, Stuttgart (1986)]
51. Nielsen, J.: The structure of periodic surface transformations. Translation by John Stillwell of : Die Struktur periodischer Transformationen von Flächen. [Math.-fys. Medd. Danske Vid.Selsk, **15**, nr.1 (1937) 77p.], [Collected Math. Papers, **2**, Birkhäuser, Boston, Basel, Stuttgart (1986)]
52. Nielsen, J.: Mapping classes of finite order. Translation by John Stillwell of: Abbildungsklassen endlicher Ordnung. [Acta Math., **75**, 23–115 (1942)], [Collected Math. Papers, **2**, Birkhäuser, Boston, Basel, Stuttgart (1986)]
53. Nielsen, J.: Surface transformation classes of algebraically finite type. Mat.-fys. Medd. Danske Vid.Selsk., **21** nr. 2 (1944), [Collected Math. Papers, **2**, Birkhäuser, Boston, Basel, Stuttgart (1986)]
54. Papakyriakopoulos, C.D.: On Dehn's lemma and the asphericity of knots. Ann. Math. **66**, 1–26 (1957)
55. Pichon, A.: Fibrations sur le cercle et surfaces complexes. Ann. Inst. Fourier (Grenoble) **51**(2), 337–374 (2001)
56. Reid, M.: Problems on pencils of small genus. unpublished note (1990) in his home page http://www.warwick.ac.uk/~masda/, click Surfaces, then Problem in geography of surfaces.
57. Satake, I.: On a generalization of the notion of manifolds. Proc. Nat. Acad. Sci. USA, **42**, 359–363 (1956)
58. Shiga, H., Tanigawa, H.: On the Maskit coordinates of Teichmüller spaces and modular transformations. Kodai Math. J. **12**, 437–443 (1989)
59. Takamura, S.: Towards the classification of atoms of degenerations. I. Splitting criteria via configurations of singular fibers. J. Math. Soc. Jpn. **56**(1), 115–145 (2004)
60. Takamura, S.: Towards the classification of atoms of degenerations. III. Splitting Deformations of Degenerations of Complex Curves. Lecture Notes in Math. **1886**, Springer Verlag, Berlin, Heidelberg, New York (2006). (See also his series of preprints, II and IV, under the same title.)
61. Tamura, I.: Foliations and spinnable structures on manifolds. In: Actes du Colloque International du C.N.R.S. de Strasbourg. Ann. Inst. Fourier, **23**, 197–214 (1973)
62. Thurston, W.: The geometry and topology of 3-Manifolds. Preprint, Princeton University Press, Princeton, NJ (1977–79)
63. Thurston, W.: On the geometry and dynamics of diffeomorphisms of surfaces. Bull. AMS. **19**, 417–431 (1988)
64. Uematsu, K.: Numerical classification of singular fibers in genus 3 pencils. J. Math. Kyoto Univ. **39**(4), 763–782 (1999)
65. Van der Waerden, B.L.: Einfürung in die algebraische Geometrie, 2nd edn. Springer-Verlag, Berlin, Heidelberg, New York (1973) [Japanese Translation: Daisu-Kikagaku Nyūmon, (Translation by H. Maeda) Springer-Verlag-Tokyo (1991)]
66. Waldhausen, F.: Eine Klasse von 3-dimensionalen Mannigfaltigkeiten. I, II. Invent. Math. **3**, 308–333 (1967)
67. Waldhausen, F.: Eine Klasse von 3-dimensionalen Mannigfaltigkeiten. I, II. Invent. Math. **4**, 87–117 (1967)
68. Waldhausen, F.: On irreducible 3-manifolds which are sufficiently large. Ann. Math. **87**, 56–88 (1968)
69. Winkelnkemper, H.E.: Manifolds as open books: Bull. Amer. Math. Soc. **79**, 45–51 (1973)
70. Winters, G.B.: On the existence of certain families of curves. Amer. J. Math. **96**, 215–228 (1974)

Index

Y. Matsumoto and J.M. Montesinos-Amilibia, *Pseudo-periodic Maps and Degeneration of Riemann Surfaces*, Lecture Notes in Mathematics 2030,
DOI 10.1007/978-3-642-22534-5,

Editorial Policy (for the publication of monographs)

1. Lecture Notes aim to report new developments in all areas of mathematics and their applications - quickly, informally and at a high level. Mathematical texts analysing new developments in modelling and numerical simulation are welcome.
   Monograph manuscripts should be reasonably self-contained and rounded off. Thus they may, and often will, present not only results of the author but also related work by other people. They may be based on specialised lecture courses. Furthermore, the manuscripts should provide sufficient motivation, examples and applications. This clearly distinguishes Lecture Notes from journal articles or technical reports which normally are very concise. Articles intended for a journal but too long to be accepted by most journals, usually do not have this "lecture notes" character. For similar reasons it is unusual for doctoral theses to be accepted for the Lecture Notes series, though habilitation theses may be appropriate.

2. Manuscripts should be submitted either online at www.editorialmanager.com/lnm to Springer's mathematics editorial in Heidelberg, or to one of the series editors. In general, manuscripts will be sent out to 2 external referees for evaluation. If a decision cannot yet be reached on the basis of the first 2 reports, further referees may be contacted: The author will be informed of this. A final decision to publish can be made only on the basis of the complete manuscript, however a refereeing process leading to a preliminary decision can be based on a pre-final or incomplete manuscript. The strict minimum amount of material that will be considered should include a detailed outline describing the planned contents of each chapter, a bibliography and several sample chapters.
   Authors should be aware that incomplete or insufficiently close to final manuscripts almost always result in longer refereeing times and nevertheless unclear referees' recommendations, making further refereeing of a final draft necessary.
   Authors should also be aware that parallel submission of their manuscript to another publisher while under consideration for LNM will in general lead to immediate rejection.

3. Manuscripts should in general be submitted in English. Final manuscripts should contain at least 100 pages of mathematical text and should always include

   - a table of contents;
   - an informative introduction, with adequate motivation and perhaps some historical remarks: it should be accessible to a reader not intimately familiar with the topic treated;
   - a subject index: as a rule this is genuinely helpful for the reader.

   For evaluation purposes, manuscripts may be submitted in print or electronic form (print form is still preferred by most referees), in the latter case preferably as pdf- or zipped psfiles. Lecture Notes volumes are, as a rule, printed digitally from the authors' files. To ensure best results, authors are asked to use the LaTeX2e style files available from Springer's web-server at:

   ftp://ftp.springer.de/pub/tex/latex/svmonot1/ (for monographs) and
   ftp://ftp.springer.de/pub/tex/latex/svmultt1/ (for summer schools/tutorials).

Additional technical instructions, if necessary, are available on request from lnm@springer.com.

4. Careful preparation of the manuscripts will help keep production time short besides ensuring satisfactory appearance of the finished book in print and online. After acceptance of the manuscript authors will be asked to prepare the final LaTeX source files and also the corresponding dvi-, pdf- or zipped ps-file. The LaTeX source files are essential for producing the full-text online version of the book (see http://www.springerlink.com/openurl.asp?genre=journal&issn=0075-8434 for the existing online volumes of LNM). The actual production of a Lecture Notes volume takes approximately 12 weeks.

5. Authors receive a total of 50 free copies of their volume, but no royalties. They are entitled to a discount of 33.3 % on the price of Springer books purchased for their personal use, if ordering directly from Springer.

6. Commitment to publish is made by letter of intent rather than by signing a formal contract. Springer-Verlag secures the copyright for each volume. Authors are free to reuse material contained in their LNM volumes in later publications: a brief written (or e-mail) request for formal permission is sufficient.

Addresses:

Professor J.-M. Morel, CMLA,
École Normale Supérieure de Cachan,
61 Avenue du Président Wilson, 94235 Cachan Cedex, France
E-mail: morel@cmla.ens-cachan.fr

Professor B. Teissier, Institut Mathématique de Jussieu,
UMR 7586 du CNRS, Équipe "Géométrie et Dynamique",
175 rue du Chevaleret
75013 Paris, France
E-mail: teissier@math.jussieu.fr

For the "Mathematical Biosciences Subseries" of LNM:

Professor P. K. Maini, Center for Mathematical Biology,
Mathematical Institute, 24-29 St Giles,
Oxford OX1 3LP, UK
E-mail : maini@maths.ox.ac.uk

Springer, Mathematics Editorial, Tiergartenstr. 17,
69121 Heidelberg, Germany,
Tel.: +49 (6221) 487-8259

Fax: +49 (6221) 4876-8259
E-mail: lnm@springer.com